Perfect Numbers and Fibonacci Sequences

Perfect Numbers and Fibonacci Sequences

Cai Tianxin

Zhejiang University, China

Translated by

Tyler Ross

New York, USA

World Scientific

NEW JERSEY · LONDON · SINGAPORE · BEIJING · SHANGHAI · HONG KONG · TAIPEI · CHENNAI · TOKYO

Published by

World Scientific Publishing Co. Pte. Ltd.

5 Toh Tuck Link, Singapore 596224

USA office: 27 Warren Street, Suite 401-402, Hackensack, NJ 07601

UK office: 57 Shelton Street, Covent Garden, London WC2H 9HE

Library of Congress Cataloging-in-Publication Data

Names: Cai, Tianxin, 1963–   author.

Title: Perfect numbers and Fibonacci sequences / Cai Tianxin, Zhejiang University, China.

Description: New Jersey : World Scientific, [2022] | Includes bibliographical references and index.

Identifiers: LCCN 2021060410 | ISBN 9789811244070 (hardcover) |

ISBN 9789811244087 (ebook for institutions) | ISBN 9789811244094 (ebook for individuals)

Subjects: LCSH: Perfect numbers. | Fibonacci numbers.

Classification: LCC QA242 .C24 2022 | DDC 512.7/2--dc23/eng20220215

LC record available at https://lccn.loc.gov/2021060410

British Library Cataloguing-in-Publication Data

A catalogue record for this book is available from the British Library.

For any available supplementary material, please visit
https://www.worldscientific.com/worldscibooks/10.1142/12477#t=suppl

Desk Editors: Vishnu Mohan/Yumeng Liu

Typeset by Stallion Press
Email: enquiries@stallionpress.com

Sculpture of Fibonacci by Giovanni Paganucci (1863), Pisa, Camposanto Monumentale.

Preface

1

The concept of perfect numbers originates in ancient Greece, with a simple definition attributed to Pythagoras: a perfect number is a positive integer n that is equal to the sum of its proper divisors (the proper divisors are all divisors of n other than n itself); or, as an equation, a positive integer n is called a perfect number if it satisfies

$$\sum_{\substack{1 \leq d < n \\ d \mid n}} d = n.$$

The two smallest perfect numbers are 6 and 28, as we can see from the identities

$$6 = 1 + 2 + 3,$$
$$28 = 1 + 2 + 4 + 7 + 14.$$

If on the other hand a positive integer n is larger or smaller than the sum of its proper divisors, then it is called an abundant or deficient number, respectively.

It is easy to prove that there are both infinitely many abundant numbers and infinitely many deficient numbers. For example, the only proper divisor of any prime number is the number one, so every prime number is deficient. Since the proof that there are infinitely many prime numbers occurs in the mathematical literature as early as the third century BCE in Euclid's *Elements*, we conclude that also there are infinitely many deficient numbers.

The *Elements* contains another important number theoretical result concerning the perfect numbers; specifically, a sufficient condition for an even number to be perfect. According to this result, if the numbers p and $2^p - 1$ are both prime, then

$$2^{p-1}(2^p - 1)$$

is a perfect number. Prime numbers of the form $2^p - 1$ as above are called Mersenne primes, after the French priest Marin Mersenne who first systematically studied such numbers in the 17th century.

But are there finitely many or infinitely many perfect numbers? More than two and a half millenia have passed, and this question has still not yet been resolved. In fact, up until the first half of the 20th century and the birth of the modern electronic computer, only twelve perfect numbers (corresponding to only twelve Mersenne primes) had been discovered.

This brings us to the perfect number problem:

(1) Are there finitely many or infinitely many even perfect numbers? and
(2) Are there any odd perfect numbers?

The perfect number problem is the oldest outstanding problem in the history of mathematics, first formulated long before any of the more recent well-known open or only lately resolved problems in mathematics, including Diophantine m-tuples, Fermat's last theorem, the Goldbach conjecture, the Waring problem, de Polignac's conjecture (and its more famous special case, the twin prime conjecture), the Riemann hypothesis, the Poincaré conjecture, the $3x + 1$ problem, the Egyptian fractions problem, the BSD conjecture, the *abc* conjecture, and so on.

The first mathematician to explicitly ask the perfect number problem was the Greek mathematician Nicomachus in the first century CE. In his *Introduction to Arithematic*, he posed in total five conjectures concerning the perfect numbers. Among these conjectures, one was that the sufficient condition given above is also a necessary condition for an even number to be perfect, another was that every perfect number is even, and a third that there are infinitely many perfect numbers. Taken together, these conjectures describe the perfect number problem as we have stated it above.

The conjecture that Euclid's condition is necessary for any even number to be perfect was taken up again in the 11th century by the Arabic mathematician Alhazen, but he too could not prove it. Finally in 1747, the Swiss mathematician proved this conjecture while living in Berlin, linking the even perfect numbers first studied in ancient Greece in a bijective correspondence with the Mersenne primes introduced in the 17th century. From Euclid to Euler, more than 2000 years had passed to reach this conclusion, which is known today as the Euclid–Euler theorem.

The Euclid–Euler Theorem. *An even positive integer n is a perfect number if and only if it has the form*

$$n = 2^{p-1}(2^p - 1),$$

where both p and $2^p - 1$ are prime numbers.

Since the perfect numbers are few and far between, people have been introducing various generalizations of the concept since the middle ages. For example, people considers numbers that are divisible by the sum of their proper divisors, or as an equations, numbers satisfying

$$\sum_{\substack{1 \leq d < n \\ d \mid n}} d = kn$$

for some integer $k \geq 1$. Such numbers are called $(k + 1)$-multiply perfect numbers.

When $k = 1$, the 2-multiply perfect numbers are the same as the ordinary perfect numbers. For $k > 1$, such formidable mathematicians as Fibonacci, Mersenne, Descartes, and Fermat, and later Lehmer, Carmichael and others have all investigated the $(k + 1)$-multiply perfect numbers, although not all of them were able to discover any such numbers. The first mathematician known to discover a $(k+1)$-multiply perfect number with $k > 1$ was the Welsh mathematician Robert Recorde, who observed in 1557 that the number 120 is 3-multiply perfect. In 1637, the French mathematician Pierre de Fermat discovered that also 672 is also 3-multiply perfect; this was the same year in which he wrote down Fermat's last theorem.

Several years later, in 1644, Fermat also discovered a 3-multiply perfect number with eleven digits; earlier, Mersenne and Descartes had found 3-multiply perfect numbers with nine and ten digits, respectively. These three mathematicians also found multiply perfect numbers of orders other than $k = 2$. The results along these lines however have been scattered and piecemeal in comparison with the depth and elegance of the Euclid–Euler theorem linking the even perfect numbers to the Mersenne primes.

From the organization and contents of this book, the reader can see that the perfect number problem originated in the eastern Mediterranean, and subsequently flourished across the three continents of Asia, Africa, and Europe. Later, with the advent of the computer age, the perfect number problem came also to the Americas and finally spread out across the world as a landmark problem not only in mathematics but also in computation.

Indeed, during the long ages of manual calculation, a total of twelve (even) perfect numbers, or equivalently twelve Mersenne primes, were discovered through the efforts and diligence of generations of mathematicians. Between the years 1952 and 1996, people used computers to find twenty-two new Mersenne primes, making up the thirteenth through the thirty-fourth known perfect numbers. And in 1996, the American computer scientist George Woltman inaugurated the Great Internet Mersenne Prime Search (GIMPS), a distributed network computational program for the discovery of Mersenne primes. Since that time, seventeen new Mersenne primes have been found using GIMPS, corresponding to the thirty-fifth through the fifty-first even perfect numbers.

2

We turn back now some 700 years and look to the Italian peninsula in the 13th century, where there appeared in the land of the former Roman empire a new mathematical concept: the Fibonacci numbers, which form the famous Fibonacci sequence. This was the first recursively defined sequence to occupy the human imagination, and it first appeared in a book called *Liber Abaci*, written by Leonardo of Pisa, known today by the name Fibonacci. In the *Liber Abaci*, the number was posed as the rabbit problem:

> A newly born pair of rabbits of opposite sexes is placed in an enclosure at the beginning of a year. Beginning with the second month, the female gives birth to a pair of rabbits of opposite sexes every month. Each new pair also gives birth to a pair of rabbits each month, starting with their second month. Find the number of pairs of rabbits in the enclosure after one year.

The Fibonacci numbers F_n are the numbers determined by this problem, with indices n corresponding to the month in which there are F_n rabbits. The numbers F_n are determined abstractly by the recursive relation

$$\begin{cases} F_0 = 0, \\ F_1 = 1, \\ F_n = F_{n-1} + F_{n-2} \quad \text{for } (n \geq 2). \end{cases}$$

The Fibonacci sequence has many remarkable properties. For example, the Italian (but naturalized French) mathematician Giovanni Cassini discovered in 1680, while he was serving as director of the Paris Observatory, the identity

$$F_{n-1}F_{n+1} - F_n^2 = (-1)^n,$$

which has since come to be called the Cassini identity. Nearly two centuries later in 1879, the French-Belgian mathematician Eugìne Catalan provided a generalization of this identity (the Catalan identity):

$$F_n^2 - F_{n-r}F_{n+r} = (-1)^{n-r} F_r^2 \quad \text{for all } 1 \leq r \leq n,$$

or

$$F_{n-r}F_{n+r} - F_n^2 = (-1)^{n-r+1} F_r^2 \quad \text{for all } 1 \leq r \leq n.$$

When $r = 1$, this is Cassini's identity, and it can be deduced from it that adjacent pairs of Fibonacci numbers are relatively prime to one another. Later, in the second half of the 20th century, the British-based mathematician Steven Vajda, who was born in Hungary and educated in Austria, generalized this result still further (Vajda's identity):

$$F_{n+j}F_{n+k} - F_n F_{n+j+k} = (-1)^n F_j F_k.$$

In 1718, the French mathematician Abraham de Moivre (1667–1754), who spent almost his entire life in England and was also the author of the central limit theorem, of fundamental importance in probability theory, discovered the following remarkable explicit expression for F_n $(n \geq 1)$:

$$F_n = \frac{1}{\sqrt{5}} \left(\left(\frac{1 + \sqrt{5}}{2} \right)^n - \left(\frac{1 - \sqrt{5}}{2} \right)^n \right).$$

This formula is generally known today as Binet's formula, after a later French mathematician.

Using this identity, it is easy to obtain the following limiting formula:

$$\lim_{n \to \infty} \phi_n = \frac{1 + \sqrt{5}}{2},$$

where $\phi_n = \frac{F_{n+1}}{F_n}$. The value on the right-hand side is $\frac{1+\sqrt{5}}{2} = 1.618\ldots$, and its reciprocal is $\frac{2}{1+\sqrt{5}} = 0.618\ldots$; this is the famous golden ratio. In fact, the German astronomer and mathematician Johannes Kepler had discovered the existence of this limit as early as 1611. The ratio ϕ_n can also be written using continued fraction notation as $\phi_n = \underbrace{[1; 1, \ldots, 1]}_{n \text{ ones}}$.

In 1774, the French mathematician Joseph-Louis Lagrange discovered that the Fibonacci numbers also exhibit periodic behavior modulo n; for every integer n there is a smallest positive integer $\pi(n)$ called the Pisano period modulo n such that

$$F_{k+\pi(n)} \equiv F_k \pmod{n}$$

for all positive integers k. For example, it is easy to calculate $\pi(1) = 1$, $\pi(2) = 3$, $\pi(3) = 8$, $\pi(4) = 6$, $\pi(5) = 20$, $\pi(6) = 24$, $\pi(7) = 16$,

$\pi(8) = 12$, $\pi(9) = 24$, $\pi(10) = 60$, $\pi(11) = 10$, $\pi(12) = 24$. If we consider the residues modulo n, we have that the residues of F_n modulo 2 repeat as $(0, 1, 1)$ and modulo 8 as $(0, 1, 1, 2, 3, 5, 0, 5, 5, 2, 7, 1)$. The terminology Pisano period is in honor of Fibonacci, Leonardo Pisano.

In 1876, the French mathematician Edouard Lucas proved another important theorem: for any two positive integers m and n,

$$\gcd(F_m, F_n) = F_{\gcd(m,n)}.$$

It follows easily from this that F_m divides F_n if and only if m divides n, and if $\gcd(m, n) = 1$, then $F_m F_n$ divides F_{mn}. As an important special case, except for $F_4 = 3$, n is a prime number whenever F_n is prime. Also, since there are arbitrarily long sequences of consecutive composite numbers, therefore there are also arbitrarily long sequences of consecutive composite Fibonacci numbers.

In 1970, the Russian mathematician Yuri Matiyasevich proved the following theorem.

Theorem (Matiyasevich). *If F_m^2 divides F_n, then F_m divides n.*

For example, $F_4^2 = 9$ divides $F_{12} = 144$ and $F_4 = 3$ divides 12; on the other hand, the converse is false in general: 5 divides 15, but $F_5^2 = 25$ does not divide $F_{15} = 610$.

From this, Matiyasevish was able to prove at the age of twenty-three, while he was still a doctoral student at Leningrad State University (now Saint Petersburg State University), that the set $\{F_{2n}\}$ of Fibonacci numbers with even index forms a Diophantine set, and confirm an earlier hypothesis due to Julia Robinson called the *J.R. Hypothesis*. This result established a negative answer to Hilbert's tenth problem, which asked for a finite process to determine whether or not a given Diophantine equation is solvable in rational integers. This is a major and surprising application of the Fibonacci sequence.

In 1972, Edouard Zeckendorf (1901–1983), a retired Belgian doctor and military officer, published a proof of the following theorem, later named after him.

Theorem (Zeckendorf). *Every positive integer admits a unique presentation as a sum of one or more distinct pairwise nonadjacent Fibonacci numbers, where in particular we include only one of $F_1 = F_2 = 1$.*

Such a representation of a positive integer is called its Zeckendorf representation, and we present several new generalizations and variations on this result below.

In addition to the theorem described above, Lucas also defined the Lucas numbers and Lucas sequences. The Lucas numbers are defined by the recurrence

$$\begin{cases} L_0 = 2, \\ L_1 = 1, \\ L_n = L_{n-1} + L_{n-1} \quad (n \geq 2). \end{cases}$$

The Lucas numbers and the Fibonacci numbers can be regarded as twin sisters. They share many independent properties, and there are also interesting results connecting the two kinds of numbers, for example:

$$F_{n+k} + (-1)^k F_{n-k} = L_k F_n,$$
$$L_n^2 - L_{n-r} L_{n+r} = (-1)^n \cdot 5F_r^2,$$
$$L_n^2 - 5F_n^2 = (-1)^n \cdot 4.$$

In 1964, the American mathematician Leonard Carlitz proved that

(1) if m divides n, then L_m divides L_n if and only if $\frac{n}{m}$ is odd, and
(2) L_m divides F_n if and only if $2m$ divides n.

The Lucas sequences are defined as follows. Let P and Q be any two nonzero integers, and consider the quadratic polynomial $X^2 - PX + Q$. The discriminant of this polynomial is

$$D = P^2 - 4Q$$

and its roots are given by

$$\alpha, \ \beta = \frac{P \pm \sqrt{D}}{2}.$$

Note that if $D \neq 0$, then necessarily $D \equiv 0$ or $1 \pmod 4$. For $n \geq 0$, we define the sequences

$$U_n(P, Q) = \frac{\alpha^n - \beta^n}{\alpha - \beta},$$
$$V_n(P, Q) = \alpha^n + \beta^n.$$

The sequence $U(P,Q) = (U_n(P,Q))_{n\geq 0}$ is called a *Lucas sequence of the first kind* with parameters P and Q, the sequence $V(P,Q) = (V_n(P,Q))_{n\geq 0}$ a *Lucas sequence of the second kind* with parameters P and Q; for fixed P, Q, the two sequences $U(P,Q)$ and $V(P,Q)$ are called *complementary Lucas sequences*.

When $P = 1$, $Q = 1$, then $U(P,Q)$, $V(P,Q)$ are the familiar sequences of Fibonacci numbers and Lucas numbers, respectively. In other words, the Fibonacci numbers and Lucas numbers form a complementary pair of Lucas sequences.

In later history, as scientific and technological knowledge advanced, people discovered that the Fibonacci sequence has direct applications in physics, chemistry, the theory of quasicrystals, and other disciplines. For this reason, the Fibonacci Association was established in 1963 by two California mathematicians. In the same year, the scientific journal *The Fibonacci Quarterly* began publication, dedicated to research results in this area. Since 1984, the Fibonacci Association has organized bianually the International Conference on Fibonacci Numbers and their Applications.

In the first two chapters of this book, we review the history and research status of the perfect numbers; in the third and fourth chapters, we describe and discuss the Fibonacci numbers, Lucas numbers, and Lucas sequences, including some very recent research results, both our own and those of our mathematical colleagues working in various countries. In addition to the perfect number problem, which we have already discussed, there remain many outstanding mysteries concerning these topics, including whether or not there exist infinitely many Fibonacci primes or infinitely many Lucas primes.

We consider also the Narayana cow sequence, introduced in 14th century by the Indian mathematician Narayana Pandita. This sequence is

$$\begin{cases} G_0 = 0, \\ G_1 = G_2 = 1, \\ G_n = G_{n-1} + G_{n-3} \quad \text{for } (n \geq 3), \end{cases}$$

a natural variation on the Fibonacci sequence. We discuss many interesting identities and congruences concerning this sequence, corresponding to similar results for the Fibonacci numbers, including analogues of both the Cassini and Vajda identities, addition rules, and various determinations of the numbers G_n with negative indices; we also develop a sister sequence to the Narayana cow sequence, imitation the relationship between the Fibonacci numbers and Lucas numbers. However, while the Fibonacci sequence has cropped up in relation with numerous problems both within and without mathematics, we have not yet discovered any similarly widespread application for the Narayana cow sequence. There are, however, some interesting open problems and conjectures; for example, apart from $n = 0$, 1, 3, and 8, are there any other integers $n \geq 0$ such that $G_{-n} = 0$?

3

For many centuries, the perfect number problem and research into the Fibonacci numbers proceeded along their separate tracks, and although the research teams devoted to their study were large and capable, it was never suspected that there should be any common ground between the two topics; certainly there was nothing like the relationship between perfect numbers and the Mersenne primes involving the Fibonacci numbers. But in the spring of 2012, the author of this book happened to consider a certain kind of square sum perfect numbers while preparing materials for a graduate seminar in number theory. These numbers satisfy the equation

$$\sum_{\substack{1 \leq d < n \\ d \mid n}} d^2 = 3n,$$

where the coefficient 3 on the right-hand side was settled upon only after some careful consideration and investigation. In less than a week, along with two graduate students, we discovered a remarkable result, a necessary and sufficient condition for the determination of such numbers:

> *the only numbers n satisfying the above equation are numbers of the form $n = F_{2k-1}F_{2k+1}$ ($k \geq 1$) where both F_{2k-1} and F_{2k+1} are twin Fibonacci primes.*

The term twin Fibonacci primes is used here to indicate a pair of Fibonacci numbers whose indices differ by two, both of which are prime numbers. In light of Lucas's theorem, if both F_{2k-1} and F_{2k+1} are prime, then also $2k-1$ and $2k+1$ must be twin primes. The first three pairs of twin Fibonacci primes are

$$(F_3, F_5) = (2, 5),$$
$$(F_5, F_7) = (5, 13),$$
$$(F_{11}, F_{13}) = (89, 233).$$

The fourth and fifth are (F_{431}, F_{433}) and (F_{569}, F_{571}), already astronomically large. Since 2 is the only even prime number, we see that 10 is the only even square sum perfect number. The next potential square sum perfect number after those given by the twin Fibonacci primes above has at least 822878 digits. Considering that the 51st Mersenne prime, discovered by the GIMPS project, has 24862048 digits, it should be possible to search for larger square sum perfect numbers by powerful computer searches. At present however, we do not know even if there is a sixth square sum perfect number, nor can we rule out the possibility that there are infinitely many. The situation is just like the situation of the Fermat primes.

If we call the classical perfect numbers M-perfect numbers in light of their affinity with the Mersenne primes, then we can call the numbers satisfying (5.1) F-perfect on account of their connection with Fibonacci primes. The Japanese mathematician Kohji Matsumoto of Nagoya University has also proposed the designations Yin and Yang perfect numbers at the 7th China-Japan Number Theory Conference in Fukuoka in 2013; note that M and F in English also stand for male (*yang*) and female (*yin*) respectively.

We also consider the more general expression

$$\sum_{\substack{1 \le d < n \\ d \mid n}} d^a = bn.$$

where a and b are any positive integers; we prove the following result: If $a = 2$ and $b \ne 3$, or if $a \ge 3$, $b \ge 1$, then this equation has at most finitely many solutions; in particular, if $a = 2$ and $b = 1$ or 2, then it has no solutions. In other words, apart from the M-perfect numbers and F-perfect numbers, there appear to be no other interesting perfect numbers along these lines. If however we consider adding a constant term to the right-hand side, then the situation changes yet again; along with three additional graduate students, we considered this more general case; and, by extending the research of perfect numbers to the consideration of affine square perfect numbers, we were able to discover connections with and interrelations between the Fibonacci numbers, Lucas numbers, and Lucas sequences, and the twin prime conjecture, the de Polignac conjecture, and Sophie Germain primes.

The de Polignac conjecture that we have now touched upon more than once states that for every integer $k \geq 1$, there are infinitely many pairs of consecutive primes with difference $2k$. When $k = 1$, this is the famous twin prime conjecture. De Polignac was born in the same year as the famous German mathematician Bernhard Riemann, and died three years earlier than him; he proposed the conjecture that bears his name in 1849.

We prove the following results along these lines: let A and k be any positive integers, and consider the equation

$$\sum_{\substack{1 \leq d < n \\ d \mid n}} d^2 = An + (k^2 + 1).$$

Then

(1) if $A \neq 2$ or k is odd, then this equation has only finitely many solutions, and
(2) if $A = 2$ and k is even, then except for finitely many computable solutions in the range $n \leq (|A| + k^2 + 1)^3$, all solutions are of the form $n = p(p + k)$ with both p and $p + k$ prime.

Corollary 1. *The equation*

$$\sum_{\substack{1 \leq d < n \\ d \mid n}} d^2 = 2n + 4k^2 + 1$$

has infinitely many solutions for every $k \geq 1$ if and only if de Polignac's conjecture holds.

Considering only the case $k = 1$, we have the following special case.

Corollary 2. *The equation*

$$\sum_{\substack{1 \le d < n \\ d \mid n}} d^2 = 2n + 5$$

has infinitely many solutions if and only if the twin prime conjecture holds.

More generally, for any integers A and B, we consider the equation

$$\sum_{1 \le d < n} d^2 = An + B$$

and prove the following result. Let P take values in integers. Except for finitely many computable solutions in the range $n \le (|A| + |B|)^3$, all integer solutions of this equation have one of the following forms:

(1) $n = U_{2k-1}(P, -1)U_{2k+1}(P, -1)$ with $A = P^2 + 2$, $B = -P^2 + 1$, and both $U_{2k-1}(P, -1)$ and $U_{2k+1}(P, -1)$ prime,
(2) $n = U_{2k}(P, -1)U_{2k+2}(P, -1)$ with $A = P^2 + 2$, $B = P^2 + 1$, and both $U_{2k}(P, -1)$ and $U_{2k+2}(P, -1)$ prime, or
(3) $n = U_{k-1}(P, 1)U_{k+1}(P, 1)$ with $A = P^2 - 2$, $B = P^2 + 1$, and both $U_{k-1}(P, 1)$ and $U_{k+1}(P, 1)$ prime.

Of course, there remain unresolved questions concerning the square sum perfect numbers and affine square sum perfect numbers. Considering de Polignac's conjecture and its generalization by Dickson (Dickson's conjecture), and, in turn, the generalization of its generalization by Sierpinski and Schinzel, we would like to know if there are equivalent or similar generalizations of the results above. Or, is there a quadratic polynomial $f \in \mathbb{Z}[x]$ such that the solutions of

$$\sum_{\substack{1 \le d < n \\ d \mid n}} d^2 = f(n)$$

can be expressed in some interesting way with reference to the prime numbers (excluding perhaps finitely many computable solutions, as above).

4

In addition to the square sum perfect number problem and its generalization, we also investigate a new type of equation with connections to the Fibonacci sequence. Research into this equation has since shown to involve rich research methodology, and the difficulty of its solution is inestimable. Even if a solution is found, it will still be worthwhile to consider the number and structure of its solutions.

In early 2013, the author casually and capriciously introduced the *abcd* equation, defined below, not expecting that it would prove to have deep connections with the Fibonacci numbers.

Definition. Let n take values among positive integer, and a, b, c, d among the positive rational numbers. The *abcd* equation is

$$n = (a + b)(c + d), \tag{1}$$

where a, b, c, d are required to satisfy $abcd = 1$.

We find that if $n = F_{2k-3}F_{2k+3}$ for some $k \geq 1$, then the *abcd* equation admits the solution $a = F_{2k-1}$, $b = \frac{1}{a}$, $c = F_{2k+1}$, $d = \frac{1}{c}$. It follows immediately that there are infinitely many integers n such that the *abcd* equation admits solutions.

Using the Pisano period property of the Fibonacci numbers, we also get the following result:

> *if n is an odd integer such that*
>
> $$n = \left(a + \frac{1}{a}\right)\left(b + \frac{1}{b}\right)$$
>
> *for some integers a, b, then necessarily $n \equiv 5 \pmod 8$; If n is an even integer such that this equation has solutions in integers, then necessarily $n = 4m$ for some $m \equiv 1 \pmod{16}$.*

The research tools and methodologies that we have had need to use to investigate this equation include not only classical arithmetic and various results involving the Fibonacci and Lucas numbers, but also the theory of elliptic curves, including the group structure and theory of torsion points of elliptic curves. We illustrate below how we have used these tools to identify and count solutions to the *abcd* equation for given n.

Note first that the *abcd* equation has solutions for fixed n if and only if the equation

$$n = x + \frac{1}{x} + y + \frac{1}{y}$$

has solutions in positive rational numbers. If $n \geq 5$, this equation admits solutions if and only if the elliptic curve

$$E_n : Y^2 = X^3 + (n^2 - 8)X^2 + 16X$$

has rational points with $X < 0$.

Example 1. The equation

$$5 = x + \frac{1}{x} + y + \frac{1}{y}$$

has unique positive rational solution $x = y = 2$. We can see this by identifying rational points with $X < 0$ of the elliptic curve

$$E_5 : Y^2 = X^3 + 25X^2 + 16X. \tag{2}$$

Using the Magma Computational Algebra System (Magma), we find that the rank of E_5 is zero. By the famous Mordell's theorem, the rational points of E_5 form a finitely generated abelian group $E_5(\mathbb{Q})$ satisfying

$$E_5(\mathbb{Q}) \cong E_5(\mathbb{Q})_{\text{tor}} \oplus \mathbb{Q}^r,$$

where $E_5(\mathbb{Q})_{\text{tor}}$ is the torsion part of $E_5(\mathbb{Q})$. We can obtain all the rational points by computation; they are $(-16, 0)$, $(-4, -12)$, $(-4, 12)$, $(-1, 0)$, $(0, 1)$, $(4, -20)$, $(4, 20)$ and the point at infinity. Substituting these into the equation above, we see that the only solution is $x = y = 2$.

For general n, we can obtain the following result using the properties of elliptic curves and the theory of torsion points:

> *if the abcd equation has any positive rational solution for some $n \geq 6$, then it has infinitely many.*

Example 2. The equation

$$13 = x + \frac{1}{x} + y + \frac{1}{y} \tag{3}$$

has infinitely many positive rational solutions. We consider the elliptic curve

$$E_{13}: \; Y^2 = X^3 + 161X^2 + 16X,$$

when $X < 0$. Magma finds that the rank of this elliptic curve is 1; via Mordell's theorem, we find a generator $P(X, Y) = (-100, 780)$, and using the group law, we conclude that every rational point of this elliptic curve with $X < 0$ has form $[2k + 1]P$ with k a nonnegative integer. We calculate (for details, see [2]) the first positive rational solution to the equation above as $(\frac{2}{5}, 10)$; the second and third are given by

$$(x, y) = \left(\frac{924169}{228730}, \frac{1347965}{156818} \right),$$

$$(x, y) = \left(\frac{33896240819350898}{3149745790659725}, \frac{12489591059767450}{8548281631402489} \right).$$

By the result we have just discussed, it follows moreover that there are infinitely many more.

The great physicist Albert Einstein (1879–1955) wrote in his autobiographical notes that "the true laws cannot be linear, nor can they be derived from linearity...". This claim was perhaps the product of excessive exuberance following his discovery of the mass-energy conversion formula of the special theory of relativity, but it has proven valid as a description of the results in this book.

Tianxin Cai
West Brook, Hangzhou
China
Spring 2022

Contents

Chapter 1

The History of Perfect Numbers

Perfect numbers, like perfect men, are very rare.

René Descartes

1.1. What are Perfect Numbers?

In the year 2000, the Clay Institute of Mathematics, with head-quarters in the state of New Hampshire in the United States, proposed a set of seven Millennium Prize Problems in Mathematics, with an award of one million US dollars for the solution of any of the problems. Landon Clay (1926–2017) was a Boston businessman who founded the Clay Mathematics Institute with the stated purpose of "increasing and disseminating mathematical knowledge". After only three years, the Poincaré conjecture became the first among the seven problems to be resolved, by the Russian mathematician Grigori Perelman (1966–). After another three years, at the International Congress of Mathematicians in Madrid, the International Mathematics Union designated Perelman the recipient of the highest honor in mathematics, the Fields Medal. Perelman, however, felt that he had already obtained sufficient satisfaction from his efforts in mathematical research, and declined the award. Some time later, he also rejected the substantial award money associated with the Millenium Prize Problems.

Also in the year 2000, the Italian mathematician Piergiorgio Odifreddi (1950–), recipient of the Galileo prize and the Peano Prize, published *The Mathematical Century: The 30 Greatest Problems of the Last 100 Years*, in which he explicates the developments and breakthroughs in 30 problems in mathematics in the course of the 20th century, including 15 in pure mathematics, 10 in applied mathematics, and 5 in mathematics and computing. Finally, he discusses four difficult unsolved problems. The first of these is the problem of perfect numbers, the remaining three are the Riemann hypothesis, the Poincaré conjecture, and the P vs. NP problem. Oddifreddi taught at the University of Milan and Cornell University, and currently holds a position as a professor of mathematical logic at the University of Turin. He has also been outspoken in his philosophical and political views, which resemble those of Bertrand Russell (1872–1970) and Noam Chomsky (1928–).

The term *perfect number* (taken from the Greek τέλειος ἀριθμός, meaning a perfect, ideal, or complete number) refers to a positive integer such that the sum of its proper divisors is exactly equal to itself. It is possible that the ancient Egyptians already took an interest in such numbers, but this cannot be confirmed. In any case, it is believed that in the sixth century BCE the ancient Greek mathematician Pythagoras (ca. 580–550 BCE) carried out research into the perfect numbers (Fig. 1.1). In particular, he knew that both 6 and

Figure 1.1.   Statue of Pythagoras

28 are perfect numbers, since

$$6 = 1 + 2 + 3,$$
$$28 = 1 + 2 + 4 + 7 + 14,$$

and he is alleged to have said "The number 6 is a perfect marriage, and health, and beauty, since its parts are complete and their sum is equal to itself."

It follows from the definition, that a natural number n is a perfect number if and only if it satisfies the equation

$$\sum_{\substack{d<n \\ d|n}} d = n, \tag{1.1}$$

where $\sum$ is the summation symbol. If we introduce the Greek letter σ (sigma) to denote

$$\sigma(n) = \sum_{d|n} d,$$

then (1.1) is equivalent to

$$\sigma(n) = 2n. \tag{1.2}$$

It is notable that in the *Book of Genesis* in the *Bible*, God is said to have created the world in six days, with the seventh day set aside for rest. The Greek-speaking Jewish philosopher Philo Judaeus, or Philo of Alexandria (ca. 15 BCE–50 CE) wrote in his book *On Creation* that the world was created in six days, and that the moon takes 28 days to revolve around the Earth. Some years later, the Greek theologian and biblical scholar Origen (ca. 185–254) and the learned ascetic Didymus the Blind (ca. 313–398) observed in addition that there are only four perfect numbers smaller than ten thousand.

At the beginning of the fifth century, the ancient Roman philosopher Saint Augustine (ca. 354–430) (Fig. 1.2) wrote in his famous *City of God* that, "Six is a number perfect in itself, and not because God created the world in six days; rather the contrary is true: God created the world in six days because this number is perfect."

Subsequently, perfect numbers, and especially the number six, have retained a special significance and appeal to human beings. For

Figure 1.2.   Statue of St. Augustine

example, the 19th century American poet J.G. Saxe (1816–1887) retold and popularized an ancient Indian parable in his poem *The Blind Men and the Elephant*, which begins with the lines

> It was six men of Indostan,
> To learning much inclined,
> Who went to see the Elephant
> (Though all of them were blind)
> That each by observation
> Might satisfy his mind. . .

They concluded one by one in turn that an elephant is like a wall (having touched its body), or a spear (having touched its tusk), or a snake (its trunk), its tree (the leg), a fan (its ear), and finally a rope (having grasped its tail).

The British writer W. Somerset Maugham (1874–1965) published a novel called the *Moon and Sixpence*, based in part on the life of the painter Paul Gauguin (1848–1903); it is regarded as a modern classic on the themes of escapism and artistry. The number six also features in the great and thought-provoking film *2001: A Space Odyssey*, made by the American director Stanley Kubrick (1928–1999) in 1968. The film is adapted from a science fiction novel and depicts the endeavor by humans to land on Jupiter some eighty billion kilometers away. There are six main characters: the captain David, the pilot Frank, the robot HAL 9000, and three hibernating astronauts.

1.2. Euclid's *Elements*

There are two well-known Euclids in the history of ancient Greece. The first is Euclid of Magara, a student of Socrates in the late fifth century BCE and founder of the Magara school of philosophy (Fig. 1.3). The second is the mathematician Euclid of Alexandria, referred to ever since as the father of geometry. Although we do not know exactly the dates of his birth and death, he was likely born in the later years of the third century BCE and lived well into the fourth. It is certain at least that he had studied for a time at Plato's Academy, and later taught mathematics at Alexandria.

Euclid's celebrated masterpiece *The Elements* deals mainly with geometry, but Books 7–9 are concerned with arithmetic and number theory, and include the definition of perfect numbers. Euclid establishes a sufficient condition for an even number to be a perfect number, namely: if p and $2^p - 1$ are both prime numbers, then

$$2^{p-1}(2^p - 1) \qquad (1.3)$$

is a perfect number (Fig. 1.4).

In order to prove this, we first determine an explicit formula for the computation of the arithmetical function $\sigma(n)$ defined in the previous section, and simultaneously prove that it is a multiplicative function, or in other words: whenever two integers m and n are

Figure 1.3. 19th century statue of Euclid by Joseph Durham in the Oxford University Museum

relatively prime, then

$$\sigma(mn) = \sigma(m)\sigma(n).$$

First, if $n = p$ is a prime number, then evidently its only divisors are 1 and p, so

$$\sigma(n) = \sigma(p) = 1 + p.$$

If $n = p^k$ is a prime power, then every divisor of n is of the form p^j with $0 \leq j \leq k$. From the summation formula for a geometric series,

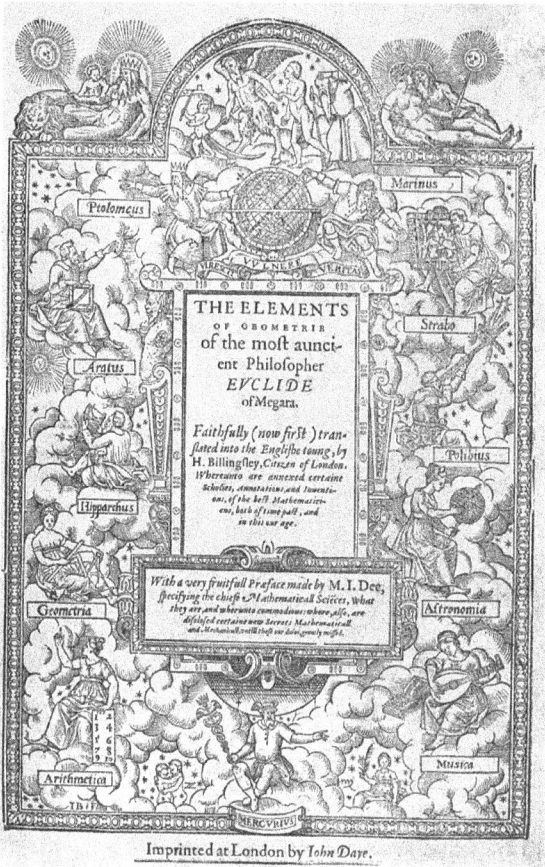

Figure 1.4.   English edition of Euclid's *Elements* (1570)

this gives

$$\sigma(n) = \sigma(p^k) = 1 + p + \cdots + p^k = \frac{p^{k+1} - 1}{p - 1}.$$

Next, if $n = pq$ has two distinct prime factors, then the divisors of n are exactly, 1, p, q, and pq, and therefore

$$\sigma(n) = \sigma(pq) = 1 + p + q + pq = (1 + p)(1 + q) = \sigma(p)\sigma(q).$$

Now suppose finally that m and n are two relatively prime positive integers. If $d|mn$, then by the nature of divisibility in number theory there are unique integers d_m and d_n such that $d_m|m$, $d_n|n$, and

$d = d_m d_n$. In fact, we can take $d_m = \gcd(d, m)$, $d_n = \gcd(d, n)$. Conversely, if $d_m | m$, $d_n | n$, then since $\gcd(m, n) = 1$, $d_m d_n | mn$. Therefore

$$\sigma(mn) = \sum_{d|mn} d = \sum_{\substack{d_m|m \\ d_n|n}} d_m d_n = \sum_{d_m|m} d_m \sum_{d_n|n} d_n = \sigma(m)\sigma(n),$$

which proves that $\sigma(n)$ is multiplicative.

Using the multiplicative property of $\sigma(n)$, we obtain an explicit formula for its computation

$$\sigma(n) = \prod_{p^k||n} \frac{p^{k+1} - 1}{p - 1},$$

where $p^k || n$ means that p^k divides n and p^{k+1} does not divide n. With this formula in hand, we can prove that numbers of the form (1.3) satisfy the identity (1.2).

Euclid's Proof. Since 2^{p-1} and $2^p - 1$ are relatively prime, it follows from the multiplicative property of $\sigma(n)$ that the divisor sum of $n = 2^{p-1}(2^p - 1)$ is

$$\sigma(n) = (1 + 2 + \cdots + 2^{p-1})(1 + 2^p - 1) = 2^p(2^p - 1) = 2n. \qquad \square$$

The sufficient condition just derived for a number to be a perfect number appears with its proof as the final proposition (Proposition 36) in Book 9 of Euclid's *Elements*. In Proposition 20 of the same book, Euclid proves that there are infinitely many prime numbers. Legend has it however that this sufficient condition was known as early as the fourth century BCE to Archytas, a member of the Pythagorean school and a close friend of the philosopher Plato (ca. 437–347 BCE) (Fig. 1.5). Archytas is considered the founder of mathematical mechanics. He also served as a military commander and is believed to have first invented the kite.

It is worth mentioning that the first Chinese translation of the *Elements* was published in 1607, translated in collaboration between the Italian missionary and sinologist Matteo Ricci (1552–1610) and the Ming scholar Xu Guangqi (1562–1633). Unfortunately, they translated only the first six books of the *Elements*, and a full edition did not appear in China until 1857, when the remaining nine books

Figure 1.5.   Archytas, disciple of Plato

were translated by the British missionary Alexander Wiley (1815–1887) and the Qing Dynasty mathematician Li Shanlan (1811–1882). So it was only at that time that perfect numbers became known in China. It is said that Xu Guangqi was a bit dissatisfied when he completed the translation of the first six books, and would have liked to continue on up to the ninth book, but Ricci did not agree. If not for this, our ancestors would have been familiar with the topics of perfect numbers and prime numbers four centuries earlier.

1.3.   Nicomachus

The perfect numbers have been from the start a source of magical attraction for mathematicians and amateurs alike, who have sought them out like gold. The third and fourth perfect numbers were discovered next, respectively, 496 and 8128. The earliest written record of their discovery is from Nicomachus of Gerasa (ca. 60 BCE–120 CE), a member of the New Pythagorean School, who mentions them in his masterpiece *Introduction to Arithmetic*, written around 100 CE (Fig. 1.6).

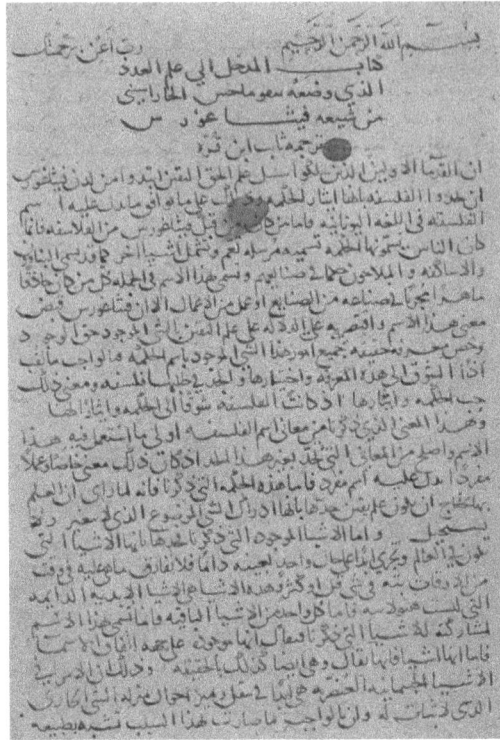

Figure 1.6.   Arabic translation of *Introduction to Arithmetic*, translated by the Syrian mathematician Thābit ibn Qurra (901), now in the British Museum

Also in the *Introduction to Arithmetic*, Nicomachus proposed five conjectures about perfect numbers. These are the earliest such conjectures. They are as follows:

(1) the nth perfect number has n digits,
(2) every perfect number is even,
(3) the perfect numbers alternatingly end with the digits 6 and 8,
(4) the sufficient condition given in Euclid's *Elements* is also a necessary condition for an even number to be perfect, and
(5) there are infinitely many perfect numbers.

Among these, the first and third conjectures were later proved to be wrong, and the fourth conjecture was proved by the 18th century Swiss mathematician Leonhard Euler (1707–1783). The second

Figure 1.7.   Jerash

and fifth conjectures comprise the problem of perfect numbers that persists through to the present day.

Nicomachus was born in Gerasa, in the Roman province of Syria, for which reason he is referred to as Nicomachus of Gerasa. In fact, around the fourth century BCE, there was also a painter named Nicomachus in Ancient Greece, but I have never had the opportunity to see his works, nor do I know if he is in any way related to Nicomachus of Gerasa. The Roman engineer, architect, writer, and connoisseur Marcus Vitruvius Polio (Vitruvius, ca. 80–70 BCE–ca. 15 AD), remembered today for his treatise *The Ten Books on Architecture*, said of this painter: "If Nicomachus has earned less fame than his contemporaries, it is only due to fortune, and not a lack of talent." In any case, Nicomachus is a name worthy of our recognition.

Gerasa is known today as Jerash, in the Kingdom of Jordan (Fig. 1.7). The city is situated between Damascus, the modern capital of Syria, and the capital Amman of Jordan and the Dead Sea.

The ruins of Gerasa comprise one of the best preserved ancient Roman cities in Jordan and are designated a UNESCO World Heritage Site in the cultural category. I travelled from Damascus to Amman in the summer of 2004, and passed through this city. I saw that the stone paved streets of the ancient city ruins were long and flat, with a forest of pillars along either side; I saw the temple on the mountain, a spectacular theater in the shape of a fan, broken baths and fountains. It is no wonder that it has been called the Pompei of Asia, also the Pompei of the Middle East.

The province of Syria was one of thirteen provinces laid out in the Roman administrative system for occupied foreign territories, the first of which was Sicilia. This system of governance remained in place for more than seven centuries. Included in the province of Syria were modern Israel, Lebanon, Syria, and Jordan. There were several flourishing commercial ports, well-developed bronze and glass industries, and a greater number of tall buildings than Rome itself; it could reasonably be described as the most prosperous of all the Roman provinces. Gerasa was furthermore an academic center, where the influence of Aristotle[a] held sway, with strong traditions of medicine, rhetoric, and law. Two of the most important and famous Roman jurists and contributors to the digest of Roman law, Papinian and Ulpian, were from the region. It was in this rich intellectual environment that the mathematician Nicomachus was nurtured.

The Introduction to Arithmetic is not like Euclid's *Elements*, in which the statements and proofs of the various propositions are given only abstractly. Nicomachus presents his results by way of many specific numerical examples. This book also includes the earliest known multiplication table from ancient Greece, and it has been regarded as an established classic for millenia. Unfortunately, no likeness has survived through the ages of its author (the mathematician and historian Morris Kline speculated in his book *Mathematical Thought from Ancient to Modern Times* that he might have been Arabic). The Latin translation of *Introduction to Arithmetic* by Lucius Apuleius Madaurensis (ca. 124–170) has

---

[a]Nicomachus was also the name of Aristotle's father and son. In the famous painting *The School of Athens*, by Renaissance painter Raphael, Aristotle is depicted holding his work *The Nicomachean Ethics* in his hands. The title of this book was in dedication to his son, and it contains also a section devoted to his father.

also been lost to history. Luckily, another Latin version, made by Boethius (ca. 477–524) survived and was used as a textbook until as late as the Renaissance.

It is worth mentioning in passing that Apuleius was a talented writer, who is most famous for the picaresque novel *The Golden Ass*, otherwise known as *Metamorphoses*, in which is related the story of a young man who accidentally turns himself into a donkey in the course of his magical experimentation. Boethius was a Roman senator, consul, administrator, and in the end a philosopher. By that time, Rome had fallen under the rule of Odoacer of the Goths, but Boethius was retained for a time under the new king, who appreciated his talents. This situation was not to last, however, and in the year 523 Boethius was framed for conspiracy to treason and thrown into prison. He was executed in secret the following year. During his imprisonment, Boethius wrote *The Consolations of Philosophy*, a long beloved masterpiece concerning the meaning of life and the ephemerality of human fate.

Since Pythagoras did not leave behind any written works, and Euclid's *Elements* devotes only a few of its books to number theory, the *Introduction to Arithmetic* can rightfully be called the first work dedicated in its entirety to arithmetic or number theory. Since this work still remains somewhat poorly known, we devote here a bit more time to describing in some detail its contents and character. The *Introduction to Arithmetic* discusses not only even numbers, odd numbers, rectangular numbers, and polygonal numbers, but also prime numbers, composite numbers, and the hexahedral numbers of the form $n^2(n+1)$.

For example, adding together the $(n-1)$th triangular number with the nth k-agonal number gives the nth $(k+1)$-agonal number. Also, the nth triangular number, the nth square number, the nth pentagonal number, and so on, for an arithmetic sequence with difference given by the $(n-1)$th triangular number. For another example, the $(n-1)$th triangular number plus the nth square number is equal to the nth pentagonal number, or as an equation,

$$\frac{n(n-1)}{2} + n^2 = \frac{n(3n-1)}{2}.$$

The American historian of mathematics Morris Kline (1908–1992) believes that from a historical perspective the influence of the *Introduction to Arithmetic* in number theory is not less than that of

Euclid's *Elements* in geometry. Both Pythagoras and Plato placed special emphasis on what later became to be known as *the quadrivium*: arithmetic, geometry, astronomy, and music. Nicomachus himself remarked that arithmetic is the mother of every other discipline. He felt that arithmetic was more primary, in the sense that without arithmetic no other science would exist, but arithmetic exists independently of any other science. His book gives a systematic, methodical, clear, and rich treatment of integers and ratios of integers without any reference to geometry. Afterwards, arithmetic overtook geometry in popularity in the Greek world, culminating in the *Arithmetica* of Diophantus.

Nicomachus also wrote a treatise on Pythagorean music theory entitled *Manual of Harmonics*, and a *Theology of Arithmetic* in two volumes, which has unfortunately survived only in fragments. Among these fragments appears a beautiful identity which is still familiar today.

Nicomachus's Theorem. The sum of the first n cubic numbers is the square of the sum of the first n positive integers (or the square of the n-th triangular number). That is,

$$1^3 + 2^3 + \cdots + n^3 = (1 + 2 + \cdots + n)^2,$$

or

$$\sum_{k=1}^{n} k^3 = \left(\sum_{k=1}^{n} k \right)^2.$$

1.4.   Sums of Squares and Cubes

Considering the condition (1.3) given by Euclid, if we put $p = 2$ or 3 we obtain the first two perfect numbers 6 and 28, respectively, and if we put $p = 5$ or 7 we get the next two perfect numbers 496 and 8128. Since it is not too difficult to check the primality of $2^p - 1$ for small primes p, it might have been expected that the third, fourth, or even the fifth perfect number would have been discovered in the time of Euclid. But this was not to be the case. Probably this was due the decline and retreat of the Greek civilization; in both the ancient Roman period and the long Middle Ages, mathematics and philosophy alike drifted into the shadows.

In the East, the fifth century Indian mathematician Aryabhata (476–550) also discovered Nicomachus's theorem. Aryabhata was the first of the major mathematicians and astronomers in Indian history. He was born near Patna, the modern capital of the state of Bihar, on the southern banks of the Ganges river. His mathematical achievements were developed mainly in the service of astronomy. In 1975, India named its first satellite in his honor. In his masterpiece the *Aryabhatiya*, Aryabhata includes the division algorithm for division with remainder (also called the Euclidean algorithm), and approximation of accurate to four decimal places, a sine table, and the above identity between the square of the sum of the first n positive integers and the sum of their cubes.

According to the *Aryabhatiya*, the author was 23 years old in the year 3600 of *Kali Yuga*. A *Yuga Cycle* in traditional Hindu cosmology is a cyclic unit of time comprising four *yugas*, *Krita* (or *Satya*) *Yuga*, *Treta Yuga*, *Dvapara Yuga*, and *Kali Yuga*, respectively the golden age, spiritual age, feminine age, and material age of mankind. The length of each age is diminished with respect to the preceding age: *Krita Yuga* lasts for 1,728,000 years, while by *Kali Yuga* the length of the Yuga has dwindled to 432,000 years. The latter cycle began in the year 3202 BCE, so year 3600 of *Kali Yuga* corresponds to the year 499 CE. By this calculation, Aryabhata was born in the year 476 CE.

We present here a concise proof of Nicomachus's theorem discovered in 1854 by Sir Charles Wheatstone (1802–1875), a British physicist and inventor who is remembered today for the Wheatstone bridge (Fig. 1.8). The Wheatstone bridge was invented by the British inventor Samuel Christie (1784–1865) in 1833. Ten years later, Wheatstone improved its design, much like Watt improved upon the steam engine, and used it to obtain the first measurement of resistance. Subsequently, this invention was widely adopted by the telegraph industry. Here is Wheatstone's proof of Nicomachus's theorem:

$$\sum_{k=1}^{n} k^3 = 1 + 8 + 27 + \cdots + n^3$$

$$= (1) + (3+5) + (7+9+11) + \cdots + ((n^2 - n + 1)$$
$$+ \cdots + (n^2 + n - 1))$$
$$= 1 + 3 + \cdots + (n^2 + n - 1).$$

Figure 1.8.   Statue of Sir Charles Wheatstone (1868)

Note that the sum of the first n consecutive odd integers is n^2, so the above identity becomes

$$\sum_{k=1}^{n} k^3 = \left(\frac{n^2+n}{2}\right)^2 = (1+2+\cdots+n)^2 = \left(\sum_{k=1}^{n} k\right)^2. \qquad \square$$

The fact that the sum of the first n odd integers is equal to n^2 can be proved by induction and was later discovered independently by a 5-year-old Russian boy named Andrei Kolmogorov (1903–1987). He developed an obsession with mathematics as a result and went on to become the greatest Russian mathematician of the 20th century, and the founder of modern probability theory. Kolmogorov was born in Tambov, about 500 kilometers southwest of Moscow, the son of an unwed mother who died in childbirth. He was first raised by two of his aunts in Tunosha near Yaroslavi at the estate of his grandfather, until, when he was seven years old, he was formally adopted by one of his aunts and together they moved to Moscow.

In order to represent the sums of the first n consecutive integers raised to powers higher than 3, it is necessary to introduce the

Bernoulli numbers and Bernoulli polynomials,

$$1^k + \cdots + n^k = \frac{B_{k+1}(n+1) - B_{k+1}}{k+1}.$$

This section has introduced just a small episode in the story of perfect numbers. In the East, for example in China, mathematicians had devoted more energy historically to practical mathematics, and few people considered problems of a more aesthetic or recreational nature, such as that of the perfect numbers. But there have been some exceptions.

1.5.   Ibn al-Haytham

In about the year 1000, another Arabic mathematician named Ibn al-Haytham (ca. 965–1040), or in its Latinate form Alhazen, having done some research into Euclid's *Elements*, put forward his own conjecture concerning the perfect numbers (Fig. 1.9). He suspected that the sufficient condition (1.3) in Euclid is also a necessary condition for an even number to be perfect, in other words:

> every even perfect number is of the form $2^{p-1}(2^p - 1)$, where both p and $2^p - 1$ are prime.

This is identical to Nicomachus's fourth conjecture, but Alhazen did not consider the odd perfect numbers, and, in any case, he too was

Figure 1.9.   Arabic mathematician Ibn al-Haytham

unable to provide a proof. Alhazen was born in Basra, a southern port city in modern Iraq, at that time a part of the Buyid emirate. It is unknown whether or not Alhazen was influenced by the work of Nicomachus, but it is perhaps relevant that Pythagoras was said to have been captured by the Persians and taken to Babylon during his sojourn in Egypt, at the time when Mesopotamia was under Persian rule.

We are reminded here too that some of the early pioneers in noneuclidean geometry also included Arabic and Persian mathematicians, for example Omar Khayyam and Nasir al-Din al-Tusi. In any case, Basra is located on the southern banks of the Shatt al-Arab, below the confluence of the Tigris and the Euphrates. It was the first city to be conquered by British forces during the Iraq war in the 21st century. The lower half of the Shatt al-Arab forms the boundary between Iraq and Iran, with Abadan, the main port city of Iran, on its north bank (Fig. 1.10).

Alhazen was also the most accomplished physicist of the Middle Ages, with contributions especially in optics, for which reason he later came to be called the father of optics. Light is of course an omnipresent and ineluctable feature of the human environment and as such an object of intellectual curiosity since ancient times. From Plato through to Claudius Ptolemy (ca. 100–170), it was believed that people see the objects around them by virtue of rays of light emitted from the eyes. Aristotle expressed some reservations as to

Figure 1.10.   Portrait of Ibn al-Haytham on the Iraqi 10 dinar banknote

this theory: why, for example, is it impossible for the eyes to see in the dark? But because of its agreement with Euclid's geometric explanations and demonstrations, the theory remained the dominant one. Alhazen corrected this and proposed instead that light was emitted by the sun and other luminous objects, and came into contact with the human eye via reflection from visible objects.

Alhazen provided also an explanation for the phenomenon in which the diameters of the sun and the moon become significantly larger as they approach the horizon. He argued that this change in visible diameter is an illusion, caused by the close proximity of the sun or the moon to the ground. Although this explanation was not immediately accepted, it gradually became the consensus opinion. He also measured the angles of incidence and refraction of light, and overturned the assertion in Ptolemy that the ratio between the two is constant. Although he stated correctly the two angles lie in the same plane, Alhazen did not discover the law of refraction, which states that the ratio of the sine of the angle of incidence to the sine of the angle of refraction is a constant given by the reciprocal of the refractive index of the medium. This law was instead first discovered experimentally by the Dutch professor of mathematics Willebrord Snellius (1580–1626), and independently demonstrated theoretically by the French mathematician René Descartes (1596–1650) in 1637.

Alhazen was honored in his lifetime as a "Second Ptolemy" and "the Physicist"; he was also known referred to as "of Basra", in honor of his hometown. The better part of his academic career was spent in Cairo, where he worked as a tutor for aristocratic families. His life there was not always entirely satisfactory. At one point he was invited by the Fatimid Caliph to contribute to the regulation and control of the flooding of the Nile. When he realized that the project required a large dam in the upper regions of the river, an entirely impracticable undertaking, he fell out of the good graces of the caliph, and legend has it that he feigned madness under fear for his life, and subsequently spent ten years under house arrest until the death of the caliph. It was during this time that Alhazen wrote his *Book of Optics* and carried out some research into number theory. It was worth noting that the Aswan Dam was indeed eventually completed — in 1970, with the help of Soviet experts.

Perhaps his most singular accomplishment occurred when Alhazen delivered a lecture at Al-Azhar University in Cairo in which he first proposed the concept of a scientific method, for which he came to also be known as the father of scientific method. Al-Azhar University opened its doors for admission officially in the year 988, and it is one of the oldest universities in the world. The Polish astronomer Johannes Hevelius (1611–1687) observed in his book *Selenography, or a Description of the Moon* remarked that Alhazen is the representative of reason, Galileo the representative of the senses. Alhazen was also a pioneer in the fields of psychophysics and experimental psychology, and the first scientist to describe the human eye on the basis of his anatomical theories. No small amount of the terminology in modern ophthamology derives from Latin translations of his work, including the words *retina, cornea, vitreous,* and others.

Another seven and a half centuries would pass before a proof was found for Alhazen's conjecture, and even today in the 21st century, Nicomachus's conjectures that there exist no odd perfect numbers and infinitely many perfect numbers remain out of reach. Some centuries after the time of Alhazen, another Egyptian mathematician named Ibn Fallus (1194–1252) listed the fifth, sixth, and seventh perfect numbers. He gave no proof of his claims however, and since he also included in his list several larger numbers which turned out not to be perfect, his work was disregarded by his Western successors. The verification of these three perfect numbers would have to wait until the 15th and 16th centuries, as we discuss in the next section.

1.6.   Mersenne Numbers and Mersenne Primes

The determination of the fifth perfect number was a long time coming, after the passage of almost thirteen and a half centuries spanning the dark period of the European Middle Ages. Finally, in the 15th century, more precisely sometime between the years 1456 and 1461, the fifth perfect number was discovered by some anonymous author as recorded in the *Survey of Arithmetic*, published by the British author Hudalrich Regius in 1536. According to our discussion of the existing literature, the previous four perfect numbers variously appeared first in Egypt or the Middle East, geographically speaking in Africa or Asia, respectively.

The fifth perfect number is the 8-digit number 33550336, corresponding to the choice of $p = 13$ in (1.3). The magnitude of this number was not enough however to disconfirm Nicomachus's first conjecture, that the nth perfect number has n digits, because it could not be verified at that time that none of the numbers between the fourth and fifth perfect numbers are perfect.

In the year 1588, the Italian mathematician Pietro Cataldi (1548–1626) discovered the sixth perfect number 8589869056 and the seventh perfect number 137438691328, corresponding to $p = 17$ and $p = 19$, respectively. At this point, the center of the universe in terms of research into perfect numbers, had decidedly relocated to Europe. Cataldi also proved that the decimal representation of any perfect number obtained from Euclid's condition must terminate in either 6 or 8. According to the discussion in the first volume of *History of the Theory of Numbers* (1919) by L.E. Dickson (1874–1954), nineteen people claimed to have discovered the sixth perfect number in the time between Nicomachus and Cataldi, and the prime number obtained in the calculation of the seventh perfect number was the largest known prime number for the following two centuries.

In the same year that Cataldi discovered the sixth and seventh perfect numbers, the French Catholic priest and mathematician Marin Mersenne (1588–1648) was born into a peasant family in Oizé, Maine county, today part of the Sarthe region west of Paris (Fig. 1.11). As a child, Mersenne studied at the church school in his hometown, and he later had the opportunity to studied Hebrew and theology in Paris after joining a religious society, before he was ordained a priest at the age of 25. When he was 32, Mersenne began to study mathematics and music, corresponding with many of the other luminaries of his age, including Descartes and Blaise Pascal (1623–1662), and scientists such as Huygens in the Netherlands and Galileo in Italy, to whose doctrine he offered his committed support.

Mersenne undertook a systematic study of numbers of the form $M_n = 2^n - 1$, with n a prime number. Such numbers are known today as *Mersenne numbers*, and, when $2^n - 1$ is also prime, as *Mersenne primes* (Fig. 1.12). We have seen already that every Mersenne prime corresponds to an even perfect number, but at that

Figure 1.11.   French mathematician Mersenne

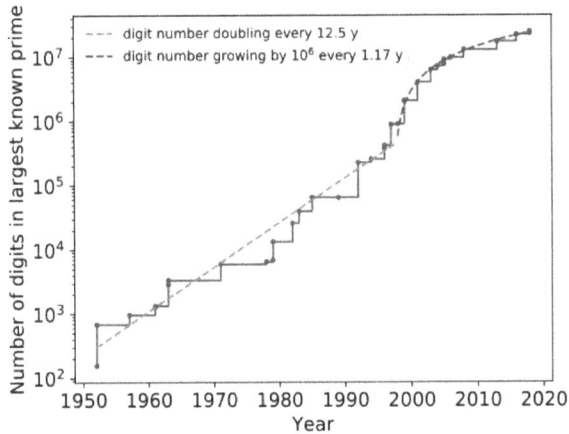

Figure 1.12.   The largest table of Mersenne prime numbers in the time of computers

time the converse proposition that every even perfect number corresponds to a Mersenne prime had not yet been proven. It was known already, however, that if n is composite, then M_n cannot be prime, or, in other words, if M_n is to be a prime number, then necessarily n must be prime.

Once people observed $M_2 = 3$, $M_3 = 7$, $M_5 = 31$, $M_7 = 127$, and $M_{13} = 8191$ are all prime, it was natural to conjecture that every number M_p with p prime is a prime number. However, Regius observed already in his *Survey of Arithmetic* that M_{11} is not prime. Rather,

$$M_{11} = 2^{11} - 1 = 2047 = 23 \times 89.$$

It is this discovery that provides the Mersenne numbers and perfect numbers with their suspense and intrigue.

In fact, the German mathematician Johannes Scheubel (1494–1570) knew already the sixth perfect number in 1557, when he included it in a footnote to his German translation of Euclid's *Elements*. This was not known until 1997, however, and so the credit for the discovery of the sixth perfect number historically has gone to Cataldi. As another aside, the word *Regius* means *royal* in Latin; today, Regius Professors are recognized at Oxford, Cambridge, the University of St. Andrews, and the Universities of Glasgow, Aberdeen, Edinburgh, and Dublin.

In addition to his substantial achievements in mathematics, Mersenne is also remembered as the father of acoustics, for his creative contributions to music theory. *Mersenne's Law* for example describes the harmonics of string vibration, and is applicable to the acoustics of the guitar and the piano. In 1635, Mersenne established the *Académie Parisienne*, an informal scientific society comprising nearly 140 correspondent members, including astronomers, philosophers, and mathematicians from France, England, Italy, and the Netherlands. He hosted numerous salons in Paris, with the Pascals (both father and son) among the frequent guests. These salons and his academy served as precursors of the Académie des Sciences, established in 1666, about two decades after Mersenne died. His death was on September 1st, of complications arising from a lung abscess. He had spent time that day with Descartes, and some scholars believe he may have drank an excess of fresh water owing to the hot weather.

In general, the largest known Mersenne prime is also the largest known prime number at any given time, so the search for Mersenne primes has been especially significant. Although it is not known whether or not there are infinitely many Mersenne primes, it is possible to use the properties of Mersenne primes to demonstrate that there are infinitely many prime numbers. Suppose for the sake of contradiction that p is the largest prime number. We can prove that any prime factor q of $2^p - 1$ must be greater than p. Indeed, since $2^p \equiv 1 \pmod{q}$, it follows that the order of 2 in the group $\mathbb{Z}/q\mathbb{Z}^\times$ of units modulo q is p. Therefore, by Lagrange's Theorem in abstract algebra, we have that p divides $q - 1$, so $p < q$, completing the contradiction.

We record here the following conjecture concerning perfect numbers and Mersenne primes.

Conjecture 1.1. *Suppose p and $2p - 1$ are both prime, and furthermore that $2^p - 1$ and $2^{2p-1} - 1$ are also prime. Then necessarily $p = 2, 3, 7$, or 31.*

We also have the following generalization of Mersenne primes. Suppose that α is a positive integer, β a nonnegative integer. Then the equation

$$\sigma(n) = 2n + 2^\alpha(2^\beta - 1)$$

has even solution $n = 2^{\alpha+\beta-1}(2^\alpha - 1)$ when $2^\alpha - 1$ is a Mersenne prime; the equation

$$\sigma(n) = 2n - 2^\alpha(2^\beta - 1)$$

has solution $n = 2^{\alpha-1}(2^{\alpha+\beta} - 1)$ when $2^{\alpha+\beta} - 1$ is a Mersenne prime.

In particular, when $\beta = 0$, this is Euclid's criterion for even perfect numbers; when $\beta = 1$, $\sigma(n)$ exceeds $2n$ by a power of 2 if and only if $n = 2^\alpha(2^\alpha - 1)$, and $2n$ exceeds $\sigma(n)$ by a power of 2 if and only if $n = 2^{\alpha-2}(2^\alpha - 1)$, where $2^\alpha - 1$ is a Mersenne prime.

Finally, we consider some variations on Mersenne primes, for example prime numbers of the form $\frac{3^p-1}{2}$ and $\frac{5^p-1}{4}$, which include 13 ($p = 3$), 1093 ($p = 7$), 797161 ($p = 13$), etc., and 31 ($p = 3$), 19531 ($p = 7$), 12207031 ($p = 11$), etc., respectively. Just like the Mersenne primes, the prime factors of such numbers must have the form $2px + 1$, but they are not necessarily congruent to ± 1 modulo 8. Furthermore, such numbers include the Wieferich prime 1093, which satisfies $2^{p-1} - 1 \equiv 1 \pmod{p^2}$; this is impossible among the Mersenne primes.

1.7. Fermat and Descartes

In the 17th century, the two great mathematicians and polymaths René Descartes and Pierre de Fermat (1601–1665) contributed tremendously to the development of human thought and civilization. They also quietly devoted some attention to the number theoretic problem of perfect numbers (Figs. 1.13 and 1.14). Descartes is responsible for the creation of the plane coordinate system and analytic geometric, a true milestone in the history of mathematical development. His philosophical doctrines of dualism and skepticism earned him the title of the father of modern philosophy according to

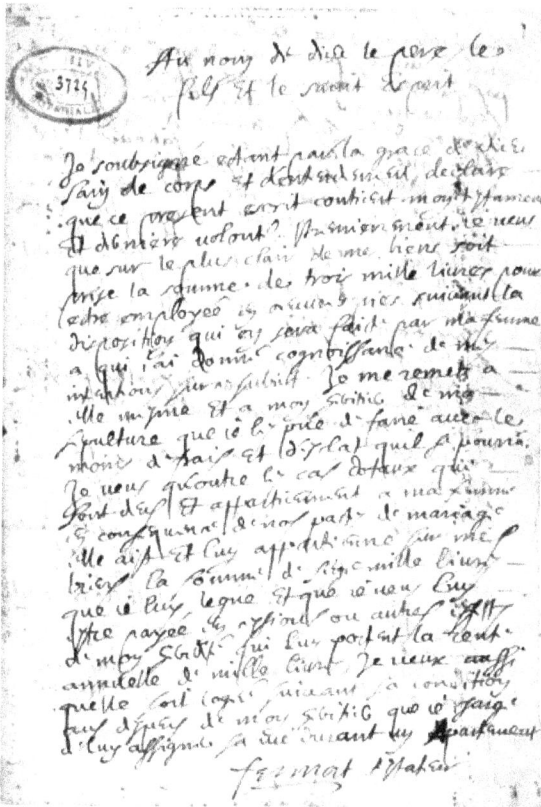

Figure 1.13.   Fermat's handwritten will, now kept at the Departmental of Haute-Garonne at Toulouse

Figure 1.14.   Descartes at work

the German idealist philosopher Hegel. Descartes made some investigations into the perfect numbers, but with little success. He quipped: "Perfect numbers, like perfect men, are rare."

In 1638, Descartes wrote in a letter to Mersenne in which he stated that he believed he could prove that every even perfect number is of the form given in Euclid, and that any odd perfect number must be the product of a prime number and the square of a different prime number. But this was only a dream, and it would be more than a century before Euler was able to see this dream through to reality. Nevertheless, as we shall describe in the following section, Descartes did manage to obtain considerable results in his investigations of the so-called k-perfect numbers.

At the same time, in the city of Toulouse in the south of France, the professional lawyer Pierre de Fermat, who was five years younger than Descartes, was devoting almost the entirety of his spare time to mathematical research. His correspondence with Pascal established probability theory as a mathematical discipline, and he discovered the fundamental principles of analytic geometry independently of Descartes. He is also generally accepted to be the inventor of differential calculus. Fermat was drawn especially to problems of number theory, and the panoply of questions he posed in that domain have kept subsequent mathematicians busy for centuries.

In the year 1640, after spending three years in the company of the perfect numbers, Fermat claimed in a letter to Mersenne that he could prove the following three propositions:

(1) if n is a composite number, then also $2^n - 1$ is composite,
(2) if n is an odd prime, then $2^n - 2$ is a multiple of $2n$, and
(3) when n is a prime number, then the prime factors of $2^n - 1$ must be of the form $2nx + 1$.

The first of these propositions is very easily dispatched: if $n = ab$ with both $a, b > 1$, then

$$2^n - 1 = (2^a - 1)((2^a)^{b-1} + \cdots + 2^a + 1)$$
$$= (2^b - 1)((2^b)^{a-1} + \cdots + 2^b + 1).$$

The second proposition is a special case of Fermat's little theorem, discussed below. The proof of the third proposition requires either the second, or Fermat's little theorem, as well as some basic results from the theory of multiplicative orders: if n is a prime number, and p is a prime factor of $2^n - 1$, then necessarily n is the order of 2 modulo p. Then from Fermat's little theorem or the second proposition above, it follows that n divides $p - 1$, and since p is odd, this implies that there must exist some positive integer x such that $p = 2nx + 1$.

Not long afterwards, Fermat wrote yet another letter to his frequent correspondent Frénicle de Bessy (ca. 1604–1674) in Paris, in which he announced the result known today as Fermat's little theorem

Fermat's Little Theorem. If p is a prime number and p does not divide an integer a, then

$$a^{p-1} \equiv 1 \ (\mathrm{mod}\, p).$$

It is easy to see that this is a generalization of the second proposition above, with the base 2 replaced by a generic integer a. We can infer from this that Fermat discovered his little theorem as a result of his research into the perfect numbers. The third proposition can be used to check whether or not $2^p - 1$ is prime for a given prime p. For example, with $p = 37$, the prime factors of $2^{37} - 1$ must be of the form $2 \cdot 37m + 1$; when $m = 1$, this gives 75, which is not prime, when $m = 2$, this gives 149, which is prime, but does not divide $2^{37} - 1$, but when $m = 3$, this gives the prime number 223, which divides $2^{37} - 1$ as $2^{37} - 1 = 223 \times 616318177$. Therefore, the prime number $p = 37$ does not generate a Mersenne prime, and equivalently does not produce a perfect number.

After he received this letter from Fermat, Mersenne was inspired to redouble his efforts, and he threw his energy into a study of the Mersenne primes and perfect numbers corresponding to primes up to $p = 257$. He published the fruits of his research in 1644, although five of his results later proved incorrect. Specifically, he mistakenly concluded that M_{67} and M_{257} are primes, although it took some 332 years before the proof was found that M_{67} is composite; he also missed the prime numbers M_{61}, M_{89}, and M_{107}. People refer to prime numbers of the form $2^p - 1$ as Mersenne primes in spite of the fact that he did not in the end discover even a single one of them in honor of his academic accomplishments and pioneering work in the research of such numbers.

As for Fermat's little theorem, Fermat himself did not provide a proof of this result, as too he did not provide proofs for so many of the results and problems that he brought into the world, which, however, has not in any way diminished his status in the history of mathematics. The term *little theorem* was first used in print to describe this result by Kurt Hensel (1861–1941) in 1913 in his *Zahlentheorie*. In Chinese this designation marks a charming contrast with *Fermat's big theorem* (费尔马大定理), but in English this famous theorem is always referred to instead as *Fermat's last theorem*, owing to the fact that the various other problems raised by Fermat were all resolved more quickly.

1.8.   The Euclid–Euler Theorem

It is now time at last for the great Swiss mathematician, physicist, and astronomer Leonhard Euler (1707–1783) to make an appearance (Figs. 1.15 and 1.16). The scope of Euler's mathematical research was astonishingly broad, including almost every mathematical topic known to his age. Certainly he too sustained a deep interest in number theory. In 1747, Euler, who was living in Berlin at the time, confirmed that conjecture by Nicomachus and Alhazen that every even perfect number is of the form (1.3). From the perspective of modern number theoretic machinery, the proof is not difficult.

Euler's Proof. Suppose n is an even number; so we can write $n = 2^{r-1}s$ where $r \geq 2$ and s is odd. If n is a perfect number, then $\sigma(n) =$

Figure 1.15.   Swiss mathematician Leonhard Euler

Figure 1.16.   Stamp of the former German Democratic Republic commemorating the 200th anniversary of the death of Leonhard Euler (1983)

$\sigma(2^{r-1}s) = 2^r s$. Since 2^{r-1} and s share no common factors, we have also that $\sigma(2^{r-1}s) = \sigma(2^{r-1})\sigma(s) = (2^r - 1)\sigma(s)$. Set $\sigma(s) = s + t$, where t is the sum of the proper divisors of s, $t = \sum_{\substack{d<n \\ d|n}} d$. Then $2^r s = (2^r - 1)(s + t)$, which gives $s = (2^r - 1)t$. Therefore t is a proper divisor of s; but also t is the sum of all proper divisors of s. This is possible if and only if $t = 1$, which shows that $s = 2^r - 1$ is prime. $\square$

Combining this result with the earlier result due to Euclid, we obtain the following theorem.

The Euclid–Euler Theorem. An even number n is a perfect number if and only if n is of the form

$$n = 2^{p-1}(2^p - 1),$$

where both p and $2^p - 1$ are prime.

In light of this, the situation of the even perfect numbers is relatively clear, and the existence of the even perfect numbers is identified with the existence of Mersenne primes. We can now reject by direct inspection Nicomachus's first and third conjectures, since the fifth perfect number has 8 digits, and both the fifth and sixth perfect numbers terminate in 6.

Euler was especially interested in the various issues handed down to mathematics by Fermat; he seems to have considered carefully and doggedly every one of them. This of course includes Fermat's little theorem, for which no proof had yet been found. Euler published its first proof in 1736. Much later, it was uncovered in a posthumous manuscript by the German mathematician G.W. Leibniz (1646–1716) that he had discovered the same proof as early 1683. In any case, Euler also proved a much stronger result, which has since come to be called Euler's theorem, in 1760, while he was still living in Berlin.

Euler's Theorem. Suppose $n > 1$, and $\gcd(a, n) = 1$. Then

$$a^{\phi(n)} = 1 \pmod{n}$$

where ϕ is the Euler totient function,

$$\phi(n) = \sum_{\substack{1 \le k \le n \\ \gcd(k,n)=1}} 1.$$

At that time, of course, Euler had no way of knowing that this result would prove to be deeply useful for cryptography in the 20th century. It is obvious that Fermat's little theorem is obtained immediately as a corollary to Euler's theorem; in fact, we can also work in the opposite direction: it is possible to prove Euler's theorem as a corollary to Fermat's little theorem.

Indeed, suppose $n > 1$, $\gcd(a, n) = 1$, and let p be any prime factor of n. By Fermat's little theorem, there exists some integer t_1 such that $a^{p-1} = 1 + pt_1$; from this it follows that there is also some integer t_2 such that $a^{(p-1)p} = (1 + pt_1)^p = 1 + p^2 t_2$, and in general some integer t_α such that $a^{\phi(p^\alpha)} = 1 + p^\alpha t_\alpha$. Therefore, in particular $a^{\phi(n)} \equiv 1 \pmod{p^\alpha}$ for every prime power p^α dividing n. Then a standard congruence argument shows that also $a^{\phi(n)} \equiv 1 \pmod{n}$, which is Euler's theorem.

In year 1772, Euler had already returned from Berlin to St. Petersburg. By this time he had gone blind in both eyes, but with the help of his assistant, he was able to produce the eighth perfect number 2305843008139952128, with a total of 19 digits, corresponding to $p = 31$ in (1.3). It had been 184 years since the previous largest perfect number was discovered. In other words, although the 17th century was described by the British philosopher Alfred North Whitehead (1861–1947) as the century of genius, and a great many mathematicians of the highest caliber indulged an interest in the perfect numbers during that time, not even a single new perfect number was found.

We present here the code in two widely used mathematical and computational software packages for the calculation of the first eight perfect numbers. These are *Mathematica* and *Maple*. Together with *MATLAB*, these are the three most popular mathematical software packages.

Mathematica Code and Output

```
Do[
    n=Prime[k];
    If[PrimeQ[2 ^n-1], Print[n,2 ^(n-1)(2^n-1)]],
    {k,1,11}
]
```

{2, 6}

{3, 28}
{5, 496}
{7, 8128}
{13, 33550336}
{17, 8589869056}
{19, 137438691328}
{31, 2305843008139952128}

Maple Code and Output

```
for k from 1 to 11 do
    n := ithprime[k];
    if isprime(2^n-1) then
        print([n, 2^(n-1)*(2^n-1)]);
    end if;
end do;
```

[2, 6]
[3, 28]
[5, 496]
[7, 8128]
[13, 33550336]
[17, 8589869056]
[19, 137438691328]
[31, 2305843008139952128]

1.9. Ivan Pervushin

More than a century passed. In 1883, in a remote village 150 kilometers away from Ekaterinaburg (see Fig. 1.17), east of the Ural Mountains in Asiatic Russia, a 56-year-old Eastern Orthodox priest named Ivan Pervushin (1827–1900) found the ninth perfect number (with a total of 37 digits and corresponding to $p = 61$ in (1.3)). Pervushin was born in Perm, Oblast, on the west side of the Ural mountains in the European part of Russia. During his long career as a village priest, he also established that the 12th and 23rd Fermat numbers are compound primes.

Figure 1.17.   Oil painting of 19th century Ekaterinaburg

The *Fermat numbers* are integers of the form

$$F_n = 2^{2^n} + 1$$

and comprise yet another deep number theoretic question introduced by Fermat. Fermat discovered that for $n = 0, 1, 2, 3$ and 4, the Fermat numbers $3, 5, 17, 257$, and 65537 respectively are all prime numbers and proposed for this reason the conjecture that every Fermat number is prime. In fact, since Fermat's last theorem was resolved at last by the British mathematician Andrew Wiles (1952–), the question of Fermat primes might well now be called Fermat's last problem. To date it is unknown even whether or not there is a sixth Fermat prime number, and whether or not there are infinitely many Fermat primes is an even more remote problem.

In the years 1877 and 1888, respectively, Pervushin found a prime factor for each of F_{12} and F_{23}. These are

$$2^{2^{12}} + 1 \text{ is divisible by } 7 \times 2^{14} + 1 = 114689, \text{and}$$

$$2^{2^{23}} + 1 \text{ is divisible by } 5 \times 2^{25} + 1 = 167772161.$$

A bit earlier, the French mathematician Edouard Lucas (1842–1891) had proved that M_{67} is a composite number, and thereby by to rest Mersenne's conjecture, although he was never able to produce an explicit prime factor. Lucas had been obsessed with the problem of Mersenne primes since he was fifteen years old, and, after nineteen years of difficult work, he proved also in 1876 that M_{127} is a prime number, working out his computations on this number with 77 digits by hand. He was able to do this because he had established the previous year a general method to test Mersenne numbers for primality. For the following three quarters of a century this was the largest known prime number, until it was supplanted after the advent of the digital age. It is still and probably always will be the largest prime number with primality verified by hand.

The Lucas Primality Test. Let n be the positive integer subject to the primality test. If there exists some integer a in the range $1 < a < n$ such that

$$a^{n-1} \equiv 1 \pmod{n}$$

and

$$a^{\frac{n-1}{q}} \not\equiv 1 \pmod{n}$$

for any prime divisor q of $n - 1$, then n is a prime number. If there does not exist any such integer a, then n is composite. Note that if n is a prime number, then any primitive root a modulo n satisfies the above conditions.

Example. Consider $n = 71$. Then the prime divisors of $n - 1 = 70$ are 2, 5, and 7. Set $a = 17$. Then

$$17^{70} \equiv 1 \pmod{71},$$
$$17^{35} \equiv 70 \not\equiv 1 \pmod{71},$$
$$17^{14} \equiv 25 \not\equiv 1 \pmod{71}, \text{ but}$$
$$17^{10} \equiv 1 \pmod{71}.$$

From this we cannot therefore conclude that 71 is prime. If instead we take $a = 11$, we have

$$11^{70} \equiv 1 \pmod{71},$$
$$11^{35} \equiv 70 \not\equiv 1 \pmod{71},$$
$$11^{14} \equiv 54 \not\equiv 1 \pmod{71}, \text{ and}$$
$$11^{10} \equiv 32 \not\equiv 1 \pmod{71}.$$

This confirms the primality of 71.

Figure 1.18. American mathematician Frank Nelson Cole

The next character to appear on the stage is the American mathematician Frank Nelson Cole (1861–1926), on Halloween in 1903 (Fig. 1.18). He stands at the blackboard, writes on it the number $2^{67} - 1$, and begins to calculate. Finally he arrived at the decimal representation 147,573,952,589,676,412,927. On the right side of the blackboard he wrote down a factorization of this number as 183,707,721 × 761,838,257,287. An hour had passed, and he returned to his seat to enthusiastic applause from the audience without having said a word. Later, he explained that this calculation had occupied his every Sunday across the span of three years.

Cole served for many years as secretary of the American Mathematical Society, and carried out his responsibilities enthusiastically. After his death, this Society decided to establish the Cole Prize in his honor. Today, this award is considered to be the most prestigious award for work in the disciplines of number theory and algebra. We may remark here that on the basis of the naming of the Mersenne numbers, the Cole Prize, and also the Fields Medal that the connection between the names of mathematical awards and objects and the significance of the mathematicians after whom they are named is often haphazard. As another example, the Chern Medal (in honor of 陈省身) and the Gauss Prize are awarded at the same time by the

International Congress of Mathematicians during its regular meetings once every four years.

A Russian writer and contemporary of Pervushin described once him with the following words:

> ... this is the modest unknown worker of science ... All of his spacious study is filled up with the different mathematical books, ... here are the books of famous mathematicians: Chebyshev, Legendre, Riemann; not including all modern mathematical publications, which were sent to him by Russian and foreign scientific and mathematical societies. It seemed I was not in a study of a village priest, but in a study of an old mathematics professor ... Besides being a mathematician, he is also a statistician, a meteorologist, and a correspondent.

Russia was the largest country in the world in the 19th century, with territory straddling the continents of Europe and Asia, the two regions separated by the dividing line of the Ural Mountains. Although Pervushin spent the majority of his life in Asiatic Russia, he was born in the eastern reaches of the European side of the Urals, in the town of Lysva, where his grandfather was a priest. In 1852, Pervushin graduated from the Kazan Theological Seminary in the capital of the Republic of Tatarstan. Nikolai Lobachevsky (1792–1856), the greatest mathematician in the history of Russia, and one of the founders of noneuclidean geometry, was living in the same city at the time, although he had already retired from his duties at Kazan University. Pervushin subsequently returned for a period to his hometown, before he relocated to the rural village where he would spend the next 25 years and founded a village school. His position as a priest both provided for his family and afforded him an abundance of free time in which to study mathematics. It is necessary to mention here that unlike Catholic priests, Eastern Orthodox priests were permitted to marry as long as they had not yet been promoted to the office of bishop, with the understand that they were no longer eligible to become bishops after marrying.

In 1883, the same year he discovered the ninth perfect number, Pervushin moved to a nearby town, where he published an article satirizing the local government and was promptly exiled to another village, in which he died at the age of 73. In the year 1893, in celebration of the 400th anniversary of the arrival of Christopher Columbus in America, a World Exposition was held in Chicago, and included a meeting of the International Mathematical Congress, a precursor to the International Congress of Mathematicians; Pervushin submitted a manuscript to the proceedings, but did not receive an invitation. Of course, the world was larger then than it is today, and Pervushin occupied one of its especially remote corners.

This brings us to the turn of the 20th century. In the years 1911 and 1914, respectively, a railway company employee in Colorado in the United States named Ralph Ernest Powers (1875–1952) discovered the tenth and eleventh perfect numbers, corresponding to $p = 89$ with 54 digits and $p = 107$ with 65 digits. This puts the perfect number discovered by Lucas in the twelfth position by magnitude. In 1934, Powers also determined that M_{241} is a composite number. The method used by the latter three mathematicians (beginning with Lucas) for identifying Mersenne primes was later adopted and refined in the 1930s by the American mathematician D.H. Lehmer (1905–1991), whose father, Derrick Norman Lehmer, and wife, Emma Lehmer, were also accomplished mathematicians. This method is known today as the Lucas-Lehmer primality test.

1.10.   The Lucas–Lehmer Primality Test

Lucas graduated from the École Normale Supérieure de Paris. He worked afterwords at the Paris Observatory, served for a time as a soldier, and eventually became a professor of mathematics. Lucas's theorem and the Lucas congruence are important results in the theory of congruence. Lucas also defined and studied what have since come to be called the Lucas numbers, which we shall investigate in Chapter 4 of this book. In 1875, Lucas conjectured that the Diophantine equation

$$\sum_{n=1}^{N} n^2 = M^2$$

has only a single nontrivial solution, given by $N = 24$, $M = 70$. This conjecture is known as the cannonball problem (Fig. 1.19). It was only proved in 1918, using elliptic functions, and later it was shown to have some connection with bosonic string theory in 26 dimensions. Tragically, Lucas cut his arm on a broken porcelain bowl

Figure 1.19.   Pyramid of cannonballs

and died of tetanus at the age of 49. The Lucas-Lehmer primality test, abbreviated as **LLT**, is as follows:

The Lucas–Lehmer Primality Test. For any odd prime p, the Mersenne number $M_p = 2^p - 1$ is prime if and only if M_p divides S_{p-2}, where

$$\begin{cases} S_0 = 4, \\ S_k = S_{k-1}^2 - 2 \text{ for } k > 0. \end{cases}$$

The proof of this claim relies on some properties of the Lucas sequence, which will be given in Section 9 of Chapter 4.

After Powers discovered the tenth and eleventh perfect numbers, no more were discovered until the night of January 30th, 1952, exactly one day before his death in a small California town, when Raphael M. Robinson (1911–1995), a mathematics professor at the University of California at Berkeley discovered the next two — for the first time, with the aid of a computer. These were the thirteenth

Figure 1.20.   Statue of Turing in Manchester (photograph by the author)

and fourteenth perfect numbers, given by $p = 421$ and $p = 607$. The same year, Robinson went on to find the fifteenth ($p = 1279$), sixteenth ($p = 2203$), and seventeenth perfect ($p = 2281$) numbers. The search for Mersenne primes and perfect numbers had entered the computer age. Through to the present day, this search is a challenge not only in pure mathematics but also in computing.

In fact, the first attempt to discover a Mersenne prime by computer search had been made three years earlier, by the British mathematician and logician Alan Turing (1912–1954), using the Manchester Mark 1 computer developed at the Victoria University of Manchester (Fig. 1.20). This attempt, however, was unsuccessful. Turing is remembered today as the father of computer science and artificial intelligence, and the highest award in computer science, the Turing Award, is named in his honor. In 1950, Turing published an influential paper entitled *Computing Machinery and Intelligence* in which he proposed what he called the imitation game but has since come to be known as the Turing Test. He argued that if a third party could not distinguish between the responses given by a humans and artificial intelligences, then the conclusion should follow that the machines can think.

Robinson was not only the first person to use a computer to discover new perfect numbers and Mersenne primes, but also the only single person to find five such numbers. Robinson was born in a small town in California, and spent both his student days and his academic career at Berkeley. After he had completed his doctoral thesis in the field of complex analysis, Robinson became a professor in 1949. His main areas of research were in mathematical logic and set theory, but he maintained also a lifelong interest in number theory. His discovery of five new perfect numbers was facilitated by the use of the Standards Western Automatic Computer (SWAC) built in 1950 by the U.S. National Bureau of Stands in Los Angeles. His wife and former student Julia Robinson (1919–1985) was also a remarkable mathematician: she contributed to the solution of Hilbert's tenth problem, and she was the first female mathematician elected to the National Academy of Sciences, and later the first she served as the first female president of the American Mathematical Society.

There are still several mathematicians and computer scientists that deserve a mention after the role of digital computing took center stage in the story of perfect numbers. The first of these is the Swedish mathematician Hans Riesel (1929–2014), who used the first computer made in Sweden (BESK, completed in 1953) to find the eighteenth perfect number, determined to $p = 3127$ in (1.3), with 969 digits. The corresponding Mersenne prime held the record for the largest known prime number for the following four years. Riesel also introduced a variation on Mersenne numbers, subsequently called *Riesel numbers*. These are positive odd integers k with the property that

$$k \times 2^n - 1$$

is composite for all natural numbers n. When $k = 1$, the numbers $k \times 2^n - 1$ are of course identical with the Mersenne numbers. A prime number of the form $k \times 2^n - 1$ with $k > 1$ is called a *Riesel primes*.

Riesel also introduced Lucas–Lehmer–Riesel test, which requires multiple initial values, in contrast with the older Lucas–Lehmer primality test.

In 1963, the Canadian computer scientist and mathematician Donald B. Gillies (1928–1975) found three Mersenne primes (the 21st through the 23rd) in less than a month using the ILLIAC II supercomputer built by the University of Illinois. Gillies was a doctoral student of the legendary John von Neumann (1903–1957) and friends with John Nash (1928–2015). His best known work was in game theory.

Another figure worth mentioned is David Slowinski, a mathematician and software engineer based for a long time out of Seattle, who discovered seven new Mersenne primes (the 27th, 29th, and the 30th through the 34th) between the years 1979 and 1996, four of them in collaboration with others. Slowinski also worked out the 26th Mersenne prime, but it had in fact already been discovered two weeks earlier by an 18-year-old high school student in Hayward, north California named Landon Curt Noll (1960–). A year before this, Noll had collaborated with his classmate Laura Nickel to find the 25th Mersenne prime. As an adult, Noll became an amateur astronomer and participated in the politics of Sunnyvale, California, home to Silicon Valley, as a council member and as vice-mayor.

1.11.   The Great Internet Mersenne Prime Search

The year 1996 was a special one in the history of Mersenne primes and perfect numbers. It was in this year that Slowinski discovered the 34th Mersenne prime. But more significantly, the American computer scientist George Woltmann (1957–) initiated the Great Internet Mersenne Prime Search (GIMPS) the first large-scale distributed computing research project to be carried out via the internet (Fig. 1.21). Woltmann himself did not discover any new Mersenne prime. Rather he released the GIMPS software and clients into the world for mathematicians and mathematics enthusiasts alike to use for free. In November of the same year, the 35th Mersenne primev was discovered through the GIMPS project.

A new Mersenne prime was discovered each year for the next three years, for the first time in the history of the subject. After

Figure 1.21.   GIMPS Logo

a gap of a single year, the 39th Mersenne prime was discovered in 2001, and starting in 2003 new Mersenne primes were discovered each year for four consecutive years. On December 27th, 2017, an electrical engineer for FedEx Express in Tennessee named Jonathan Pace discovered the 50th prime number, corresponding to $p = 77232917$, with 23249425 digits. There is an amusing story in connection with this: in the year it was discovered, the Japanese publishing house Nanairosha published a book entitled *The Largest Prime Number of 2017*, containing nothing other this number, spread across 791 pages (Fig. 1.22). The book rose quickly to the number one position on Amazon and sold out in four days.

This brings us to the 51st, and most recently discovered Mersenne prime. This number is obtained from $p = 82589933$ and contains 24,862,048 digits, and it remains today the largest known prime number. If it were printed in a standard font size, it would stretch across more than a hundred kilometers. It was discovered in the winter of 2018 via a computer volunteered for the GIMPS project by Patrick Laroche, an IT professional based in Ocala, Florida. Since no new Mersenne primes were found in 2019, or have been found since then, the record of four consecutive years of Mersenne prime discoveries remains unbroken.

The mathematician Curtis Cooper of the University of Central Missouri has in particular had good luck with GIMPS: he found four Mersenne primes in the course of about decade; these are the 43rd, the 44th, the 48th, and the 49th in order of magnitude. His luck since then has not been quite as good, but in any case Cooper is also the longstanding editor of *Fibonacci Quarterly*.

M77232917

2017年

最大の素数

50th Mersenne prime

Figure 1.22. Cover of the Japanese book *The Largest Prime Number of 2017*

In 1999, the Electronic Frontier Foundation (EFF) announced the EFF Collaborative Computing Awards for finding large prime numbers, inspired by the GIMPS program. The award conditions stipulation that the first individual to find a prime number with at least 10,000,000 digits will receive $100,000USD, the first individual to find a prime number with at least 100,000,000 digits will receive $150,000USD, and the first individual to find a prime number with at least 1,000,000,000 digits will receive $250,000USD. The first of these was awarded on October 22, 2009 to GIMPS and the Mathematics Department at the University of California, Los Angeles, where the software was installed and maintained by computing manager Edson

Smith, for the discovery of the 47th Mersenne prime, with 12,978,189 digits. *Time* magazine included this discovery among its list of the fifty best scientific inventions of 2008. In fact, the 45th and 46th Mersenne primes are also large enough to earn the prize, but these were only found some months later. Such reversals of order have been common throughout the history of the search for Mersenne primes; for example, the 29th Mersenne prime was found in 1988, although the 30th and 31st Mersenne primes had been discovered earlier, in 1983 and 1985, respectively. At that time, there were anyway no cash awards associated with the discovery of large primes.

Prior to the age of GIMPs, Dickson devoted fifty pages in the first volume of his *History of the Theory of Numbers* to review the history of Mersenne primes and perfect numbers. He points out that the search for such numbers had inspired the efforts of such mathematical luminaries as Fibonacci, Cardano, Tartaglia, Leibniz, Catalan, Sylvester, Carmichael, and Dickson himself. Among them, Fibonacci mistakenly believed that he had discovered a method for finding infinitely many perfect numbers. Tartaglia also mistakenly believed that the numbers

$$1 + 2 + 4$$
$$1 + 2 + 4 + 8$$
$$\vdots$$

alternate between prime and composite numbers. Even Leibniz fell into the mistaken belief that the number $2^n - 1$ is prime if and only if n is prime.

The Lucas–Lehmer test provided a simple criterion for the verification of Mersenne primes, but the numbers involved in the sequence grow extremely fast. The first eight numbers in this sequence are:

  4, 14, 194, 37624, 1416317954, 2005956546822746114,
  4023861667741036022825635656102100994,
  16191462721115671781777559070120513664958590125499158514329308740
  975788034.

These eight numbers are enough to verify the primality of M_2, M_3, M_5, and M_7. This gives an idea of computational demands involved in the determination of Mersenne primes by the Lucas–Lehmer test. The modern search for new Mersenne primes and perfect numbers is

Figure 1.23.   James Maynard, a rising star of number theory

a collaboration not only between major computer companies and universities, but also between both professional and amateur mathematicians and computer scientists. Hundreds of thousands of computers from nearly 200 countries and regions around the world now participate in the GIMPS project, using more than 1.83 million central processing units for networking. But even at this scale, the Lucas–Lehmer test remains the most effective method with which to discover new Mersenne primes.

Perhaps inspired by the example of the GIMPS project, the British mathematician and 1998 Fields Medal winner Timothy Gowers (1963–) launched the Polymath Project on his blog in 2009, a collaborative online effort to make headway into some of the most important and difficult outstanding problems in mathematics. The Chinese-Australian mathematician and 2006 Fields Medal winner Terence Tao (1975–) participated heavily in the Polymath5 Project. After the Chinese-American mathematician Zhang Yitang proved that there infinitely many pairs of consecutive primes separated by a difference of no more than 70 million, Tao and other number theorists initiated the Polymath8 Project to improve this bound as part of the effort to solve the famous twin prime conjecture. In large part due to the contributions of British mathematician James Maynard (1987–), the bound was improved to 246, for which contributions Maynard was awarded the 2020 Cole Prize (Fig. 1.23).

Chapter 2

Questions Related to Perfect Numbers

> Whether or not there are any odd perfect numbers ... is a most difficult problem.
>
> Leonhard Euler

> Mathematics is the only true metaphysics.
>
> Lord Kelvin

2.1. Properties of Even Perfect Numbers

In this chapter, we discuss some properties of perfect numbers and their generalizations. First, we introduce the concept of a triangular number. A triangular number is a positive integer n such that n points can be arranged in a triangle, as for example in bowling or billiards (Fig. 2.1). The first ten triangular numbers are 1, 3, 6, 10, 15, 21, 28, 36, 45, 55. It is sometimes also convenient to consider zero a triangular number.

Property 1. All even perfect numbers are triangular numbers; that is, every even number is a sum of consecutive positive integers.

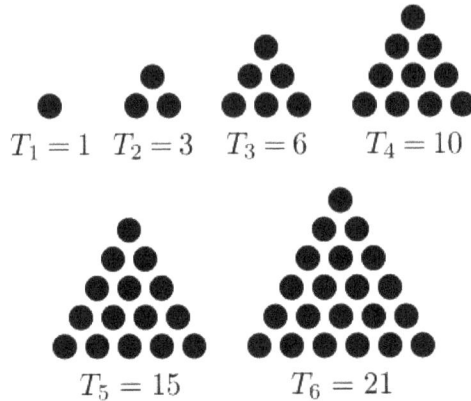

$$T_1 = 1 \quad T_2 = 3 \quad T_3 = 6 \quad T_4 = 10$$

$$T_5 = 15 \qquad T_6 = 21$$

Figure 2.1.   The first six triangular numbers

For example,

$$6 = 1 + 2 + 3 = \binom{4}{2},$$

$$28 = 1 + 2 + 4 + 7 + 14 = \binom{8}{2},$$

$$496 = 1 + 2 + \cdots + 248 = \binom{32}{2},$$

$$8128 = 1 + 2 + \cdots + 4064 = \binom{128}{2}.$$

This follows immediately from the fact that if we put $m = 2^p - 1$ for some perfect number $n = 2^{p-1}(2^p - 1)$, we get

$$n = \binom{m+1}{2}. \qquad \square$$

Property 2. Except for $n = 6$, every perfect number admits a presentation as a sum of consecutive odd cubes.

For example,

$$28 = 1^3 + 3^3$$

$$496 = 1^3 + 3^3 + 5^3 + 7^3$$

$$8128 = 1^3 + 3^3 + \cdots + 13^3 + 15^3$$

$$33550336 = 1^3 + 3^3 + \cdots + 125^3 + 127^3.$$

This follows from Nicomachus's theorem, discussed in Chapter 1:

$$1^3 + 3^3 + \cdots + (2n-1)^3 = (1^3 + 2^3 + \cdots + (2n)^3)$$
$$- (2^3 + 4^3 + \cdots + (2n)^3))$$
$$= \left(\frac{2n(2n+1)}{2}\right)^2 - 8\left(\frac{n(n+1)}{2}\right)^2$$
$$= n^2(2n^2 - 1).$$

If we substitute $n = 2^{(p-1)/2}$, we get

$$1^3 + 3^3 + \cdots + (2n-1)^3 = 2^{p-1}(2^p - 1). \qquad \square$$

Property 3. Every even perfect number admits a presentation as a sum of a prime number of consecutive powers of 2.

For example,

$$6 = 2^1 + 2^2,$$
$$28 = 2^2 + 2^3 + 2^4,$$
$$496 = 2^4 + \cdots + 2^8,$$
$$8128 = 2^6 + \cdots + 2^{12},$$
$$33550336 = 2^{12} + \cdots + 2^{24}.$$

This is a simple consequence of the familiar summation formula for geometric series:

$$2^a + \cdots + 2^{a+b} = 2^a(2^{b+1} - 1)$$

with $a = b = p - 1$. $\qquad \square$

Property 4. The final digit in the decimal representation of any even perfect number is always 6 or 8. If it is 8, then the final two digits are 28.

Proof. With $p = 2$, we have the perfect number 6, there is nothing to prove. If $p > 2$, then $p \equiv 1$ or $3 \pmod 4$. If $p \equiv 1 \pmod 4$, then $2^{p-1}(2^p - 1) \equiv 6 \pmod{10}$, and if $p \equiv 3 \pmod 4$, then $2^{p-1}(2^p - 1) \equiv 8 \pmod{10}$. Moreover, when $p \equiv 3 \pmod 4$, then $2^{p-1}(2^p - 1) \equiv 28 \pmod{100}$. $\qquad \square$

Among the 51 known perfect numbers, 31 of the corresponding prime numbers are $p \equiv 1 \pmod 4$, and 19 of the corresponding prime numbers are $p \equiv 3 \pmod 4$.

Property 5. With the exception of 6, every even perfect number is congruent to 1 modulo 9.

For example,

$$28 = 9 \times 3 + 1,$$

$$496 = 9 \times 55 + 1,$$

$$8128 = 9 \times 903 + 1.$$

Note that $2^6 \equiv 1 \pmod 9$, so

$$2^{p-1}(2^p - 1) \equiv \begin{cases} 1 \times 1 \equiv 1 \pmod 9 & \text{if } p \equiv 1 \pmod 6, \\ 7 \times 4 \equiv 1 \pmod 9 & \text{if } p \equiv 5 \pmod 6. \end{cases} \quad \square$$

Property 6. Every even perfect numbers is a pernicious number.

A *pernicious number* is a positive integer n such that 1 occurs a prime number of times in the binary representation of n. The smallest pernicious number is 3, since the binary representation of 3 is $3_{10} = 101_2$, containing a prime number $p = 2$ occurrences of the digit 1.

The even perfect numbers have the form

$$2^{p-1}(2^p - 1) = 2^{p-1}(2^{p-1} + 2^{p-2} + \cdots + 2 + 1)$$

$$= 2^{2p-2} + 2^{2p-3} + \cdots + 2^p + 2^{p-1}.$$

It follows that the binary representation of $2^{p-1}(2^p - 1)$ consists of p ones followed by $p - 1$ zeros. For example,

$$6_{10} = 110_2,$$

$$28_{10} = 11100_2,$$

$$496_{10} = 111110000_2,$$

$$8128_{10} = 1111111000000_2,$$

$$33550336_{10} = 1111111111111000000000000_2.$$

Property 7. Every even perfect number is a practical number.

A positive integer n is called a *practical number* if every positive integer smaller than n admits a presentation as a sum of divisors of n. For example, 12 is a practical number, since we have the presentations $1 = 1$, $2 = 2$, $3 = 3$, $4 = 4$, $5 = 3 + 2$, $6 = 6$, $7 = 4 + 3$, $8 = 6 + 2$, $9 = 6 + 3$, $10 = 6 + 4$, $11 = 6 + 3 + 2$ using the divisors 1, 2, 3, 4, and 6 of 12.

Proof. It is easy to see that every integer of the form $n = 2^k$ is a practical number. Now consider any even perfect number $n = 2^{p-1}(2^p - 1)$. If $1 \leq k < 2^{p-1}$, then it is also easy to see that $k(2^p - 1)$ admits a presentation as a sum of divisors of n. Now suppose

$$k(2^p - 1) < x < (k + 1)(2^p - 1) \quad \text{and} \quad 1 \leq k < 2^p - 1,$$

or equivalently,

$$0 < x - k(2^p - 1) < 2^p - 1.$$

It follows from the fact that 2^p is practical that $x - k(2^p - 1)$ admits a presentation as a sum of numbers drawn from $1, 2, \ldots, 2^{p-1}$. Moreover, $k(2^p - 1)$ admits a presentation as a sum of numbers drawn from $2^p - 1, 2(2^p - 1), \ldots, 2^{p-2}(2^p - 1)$. Adding these presentations to one another produces a presentation of x as a sum of divisors of n. $\square$

Property 8. Apart from 6, if we add together the digits of any even perfect number and iterate until there remains only a single digit, the process always terminates in 1.

For example,

$$28:\ 2 + 8 = 10, \quad 1 + 0 = 1;$$
$$496:\ 4 + 9 + 6 = 19, \quad 1 + 9 = 10, \quad 1 + 0 = 1;$$
$$8128:\ 8 + 1 + 2 + 8 = 19, \quad 1 + 9 = 10, \quad 1 + 0 = 1;$$
$$33550336:\ 3 + 3 + 5 + 5 + 0 + 3 + 3 + 6 = 28, \quad 2 + 8 = 10,$$
$$1 + 0 = 1.$$

This follows from that fact that $2^p + 1 \equiv 0 \pmod 3$ whenever p is an odd prime; therefore

$$2^{p-1}(2^p - 1) = 1 + \frac{(2^p - 1)(2^p + 1)}{2}$$
$$= 1 + 9\left(\frac{(2^p + 1)/3}{2}\right).$$

Since $2^{p-1}(2^p - 1)$ is nonzero modulo 10, therefore by a familiar argument the sum of the digits remains 1 modulo 9 until necessarily the process terminates in 1.

2.2. Open Questions

Question 2.1. Just how many even perfect numbers are there?

Through the collective efforts of many mathematicians and mathematics enthusiasts across many centuries, a total of 51 perfect numbers have so been discovered. The largest of these corresponds to the Mersenne prime $2^{82589933} - 1$, which has $24,862,048$ digits; the perfect number itself has $49,724,095$ digits, the last of which is 6, and would span a length of more than two hundred kilometers if it were printed out at a normal font size. This Mersenne prime was discovered on December 7th, 2018, and remains to this day the largest known prime number. The corresponding perfect number is naturally also the largest known perfect number.

Nevertheless, it remains unknown whether or not the number of even perfect numbers is finite or infinite.

Question 2.2. Do there exist any odd perfect numbers?

So far, all of the perfect numbers that have been found have been even. Are there any odd perfect numbers? Nobody has been able to resolve this question one way or the other. It can be shown however that if there do exist any odd numbers, they must be very large and satisfy a list of rather severe conditions, as discussed in the next section. The conjecture that every perfect number is even has a long history; indeed, this was Nicomachus's second conjecture about perfect numbers, as we have already seen. In the medieval period, the French writer, theologian, translator, and humanist Jacques Lefèvre d'Étaples (ca. 1455–1536) also promoted this theory (Fig. 2.2).

Lefèvre was a religious reformer and early precursor to the Protestant Reformation movement in Europe who sought to divorce religious studies from the esoterica of scholastic philosophy. In 1530, he completed a translation of the Bible from the Latin of the Catholic Church into vernacular French. His work had exerted a deep influence on the younger generation of scholars and reforms, including

Figure 2.2.   The French religious reformer Jacques Lefèvre

Martin Luther. In his middle age, he also published translations and annotations of Aristotle's *Ethics*, *Politics*, and *Metaphysics*. He was also interested in mathematics and physics and published many lectures on these topics. In 1496, after investigating the question of perfect numbers, he proposed that every perfect number is given by Euclid's criterion.

In any case, the problem of odd perfect numbers, first proposed in the first century by Nicomachus, remains one of the oldest open problems in mathematics. Even the introduction of powerful digital computers or powerful external hypotheses into the perfect number repertory has not facilitated any cracks in the problem. It seems possible that the story of perfect numbers, which is the longest story in mathematics, might persist into the future even if humans were to come into contact with extraterrestrial intelligences.

As for Nicomachus's third conjecture, that the even perfect numbers terminate alternately in 6 and 8, this was proven false as a corollary to the Euclid–Euler theorem, although it is true that every even perfect number terminates in one of these two digits. Since then no new conjectures have been put forward concerning the digits of perfect numbers, we furnish one such conjecture at the end of this chapter.

2.3. Odd Perfect Numbers

Around the time he was proving the full characterization of even perfect numbers, Euler also remarked "whether or not there are any odd perfect numbers... is a most difficult question". Nevertheless, Euler also proved that if any odd number is perfect, it must be of the form

$$n = p^\alpha m^2, \tag{2.1}$$

where p is an odd prime not dividing m, and $p \equiv \alpha \equiv 1 \pmod 4$. In particular, necessarily $n \equiv 1 \pmod 4$. This shows that any odd perfect number must include an odd power of a prime number in its prime factorization and therefore cannot be a square number.

The proof is actually quiet easy: suppose n is an odd perfect number with prime factorization $n = p_1^{\alpha_1} \cdots p_k^{\alpha_k}$, where $p_1, \ldots, p_k$ are odd primes. From the explicit expression for the σ function, we have

$$\sigma(n) = \prod_{j=1}^{k} \frac{p_j^{\alpha_j+1} - 1}{p - 1} = 2 p_1^{\alpha_1} \cdots p_k^{\alpha_k}. \tag{2.2}$$

Therefore, we can assume without loss of generality that

$$\sum_{t=0}^{\alpha_1} p_1^t \equiv 2 \pmod 4 \tag{2.3}$$

and

$$\sum_{t=0}^{\alpha_j} p_j^t \equiv 1 \pmod 2 \tag{2.4}$$

for $2 \le j \le k$. From (2.3) it follows that $p_1 \equiv \alpha_1 \equiv 1 \pmod 4$, and from (2.4) we can conclude that all of $\alpha_2, \ldots, \alpha_k$ are even. This completes the proof.

We prove next that if n is an odd perfect number, then m in (2.1) must have at least two prime factors; so n must have at least three prime factors. Suppose otherwise that $k = 2$ in (2.2), that is

$$\frac{p^{\alpha+1} - 1}{p - 1} \cdot \frac{q^{2\beta+1} - 1}{q - 1} = 2p^{\alpha}q^{2\beta},$$

where we have used the fact that the second factor must have an even exponent. From this it follows that

$$2 = \frac{1 - \frac{1}{p^{\alpha+1}}}{1 - \frac{1}{p}} \cdot \frac{1 - \frac{1}{q^{2\beta+1}}}{1 - \frac{1}{q}}$$

$$< \frac{p}{p - 1} \cdot \frac{q}{q - 1}$$

$$\leq \frac{3}{2} \cdot \frac{5}{4}$$

$$= \frac{15}{8},$$

which is a contradiction.

In 2003, the American mathematician Carl Pomerance (1944–) presented a heuristic argument against the existence of odd perfect numbers (Fig. 2.3). In 2007, Pace Nielsen further proved that any odd perfect number must have at least 9 distinct odd factors and at least 101 prime factors counting multiplicity; if 3 is not among them, then there must be at least 12 distinct prime factors. In 2012, Pascal

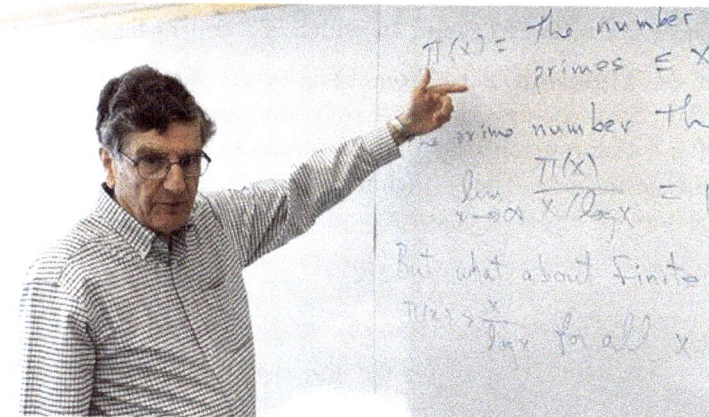

Figure 2.3. American mathematician Carl Pomerance

Ochem and Michaël Rao proved that any odd perfect number must be larger than 10^{1500}.

2.4. Touchard's Theorem

In 1953, the French mathematician Jacques Touchard used Euler's result and the multiplicative property of the σ function to prove the following theorem.

Theorem 2.1 (Touchard). *Any odd perfect number must be either of the form $12m + 1$ or of the form $36m + 9$.*

In order to prove this theorem, we need first to prove a lemma.

Lemma 2.1. *If $n \equiv -1 \pmod{6}$, then n is not a perfect number.*

Proof. The proof is by contradiction. Suppose that $n \equiv -1 \pmod{6}$ is a perfect number. Then also $n \equiv -1 \pmod{3}$. Consider any divisor d of n. If $d \equiv 1 \pmod{3}$, then $\frac{n}{d} \equiv -1 \pmod{3}$. If $d \equiv -1 \pmod{3}$, then $\frac{n}{d} \equiv 1 \pmod{3}$. That is, for every divisor d of n, $\frac{n}{d} \equiv -d \pmod{3}$. Since we have seen already that n cannot be a square number, therefore

$$\sigma(n) = \sum_{\substack{d \mid n \\ d < \sqrt{n}}} \left(d + \frac{n}{d} \right) \equiv 0 \pmod{3}.$$

But also $\sigma(n) = 2n \equiv 1 \pmod{3}$, which is a contradiction. $\square$

Proof of Theorem 2.1. From Lemma 2.1, necessarily $n \equiv 1$ or $3 \pmod{6}$. Suppose first that $n \equiv 1 \pmod{6}$. Since also $n \equiv 1 \pmod{4}$ by Euler's theorem on odd perfect numbers, it follows that $n \equiv 1 \pmod{12}$. Similarly, if $n \equiv 3 \pmod{6}$, then $n \equiv 9 \pmod{12}$, say $n = 12k + 9 = 3(4k + 3)$. If k is not a multiple of 3, then $(3, 4k + 3) = 1$, so

$$\sigma(n) = \sigma(3)\sigma(4k + 3)$$
$$= 4\sigma(4k + 3)$$
$$\equiv 0 \pmod{4}.$$

But also $\sigma(n) = 2n = 6(4k + 3) \equiv 2 \pmod{4}$, which is a contradiction. Therefore, k is a multiple of 3, so $n = 36m + 9$ where $k = 3m$.

This result was improved upon slightly by the Australian mathematician Tim Roberts. Roberts proved that if n is an odd perfect number, then it must satisfy one of $n = 12m + 1$, $n = 324m + 81$, or $n = 468m + 117$; the latter two conditions of course are special cases of $n \equiv 9 \pmod{36}$.

Proof. Suppose n is an odd perfect number and $3^k || n$. From Euler's formula (2.2), necessarily k is even. Note also that $\sigma(3^k) = 1 + 3 + \cdots + 3^k$ divides $\sigma(n) = 2n$. If $k = 0$, then Touchard's theorem proves that $n \equiv 1 \pmod{12}$. Similarly, if $k = 2$, then $n \equiv 9 \pmod{36}$; moreover since $\sigma(3^2) = 13$, $n \equiv 0 \pmod{13}$. It follows by the Chinese Remainder Theorem (秦九韶定理, Qin Jiushao's Theorem) that $n \equiv 117 \pmod{468}$. Suppose finally that $k > 2$. Then $3^4 = 81$ divides n, and since again $n \equiv 9 \pmod{36}$, therefore also $n \equiv 81 \pmod{324}$.

$\square$

We close this section with some remarks from the British mathematician James Joseph Sylvester (1814–1897), who also taught at the University of Virginia and Johns Hopkins University in the United States (Fig. 2.4). Sylvester was primarily an algebraist; he worked for a long time in collaboration with his compatriot Arthur Cayley (1821–1895) and together they established the theory of determinants and established the foundations for algebraic invariant theory.

Figure 2.4. British mathematician James Joseph Sylvester

But he was also interested in number theory, in particular the theory of partitions and Diophantine analysis. In 1888, Sylvester said of odd perfect numbers that:

> ... a prolonged meditation on the subject has satisfied me that the existence of any one such — its escape, so to say, from the complex web of conditions which hem it in on all sides — would be little short of a miracle.

2.5.   Deficient and Abundant Numbers

Since perfect numbers are so rare, we extend our study to include the deficient and abundant numbers, which were first introduced by Nicomachus.

Consider for example the number 4. Its proper divisors are 1 and 2 with sum $1 + 2 = 3$, which is less than 4. Such a number is called *deficient*. On the other hand, the proper divisors of 12 are 1, 2, 3, 4, and 6, with sum $1 + 2 + 3 + 4 + 6 = 16$, which is larger than 12. Such a number is called *abundant* (Fig. 2.5).

Every prime number is deficient, since the only proper divisor of a prime number is 1. Odd numbers with at most two distinct prime factors are also deficient, and any divisor of a deficient number or a perfect number is deficient. The smallest abundant number is 12, with prime factors 2 and 3. The smallest odd abundant number is

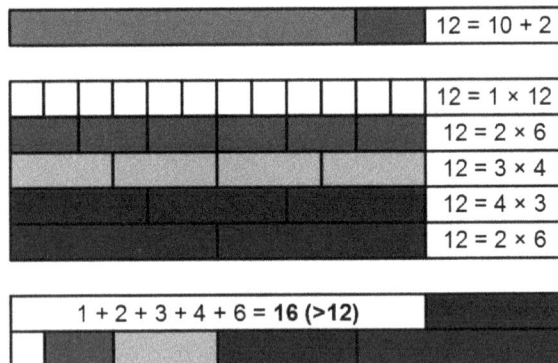

Figure 2.5.   Depiction of the abundant number 12 with colored blocks

945, with prime factors 3, 5, and 7. Finally, the smallest abundant number that is not divisible by either 2 or 3 is 5391411025, with prime factors 5, 7, 11, 13, 17, 19, 23, and 29.

If n is a perfect number, then kn is an abundant number for all $k \geq 2$. For example, if the perfect number 6 divides n, then the proper divisors of n include 1, $\frac{n}{6}$, $\frac{n}{3}$, $\frac{n}{2}$, the sum of which already is $n + 1$. Similarly, the multiples of an abundant number are also abundant; for example, 20 is an abundant number, and the divisor sum of any of its multiples $n = 20k$ includes $\frac{n}{2} + \frac{n}{4} + \frac{n}{5} + \frac{n}{10} + \frac{n}{20} = n + \frac{n}{10}$. There are both infinitely many deficient numbers, and infinitely many abundant numbers. In 1998, Marc Deléglise proved that the abundant numbers (defined by $\sigma(n) \geq 2n$ and therefore including the perfect numbers) have a natural density between 0.2474 and 0.2480. In 2006, Sándor *et al.* proved in their *Handbook of Number Theory I* that when n is large enough, there must be a deficient number in the interval $[n, n + (\log n)^2]$.

There is one especially important open question in theory of deficient and abundant numbers: apart from the positive powers of 2, are there any deficient numbers n such that $\sigma(n) = 2n - 1$ (such a number is called an *almost perfect number*); and, are there any abundant numbers n such that $\sigma(n) = 2n + 1$ (such numbers are called *quasiperfect numbers*).

If we consider instead abundant numbers n satisfying $\sigma(n) = 2n+2$, there are many; the smallest is 20, since the sum of the proper divisors of 20 is $1 + 2 + 4 + 5 + 10 = 22$. In general, if $2^n - 3$ is prime, then $2^{n-1}(2^n - 3)$ satisfies this condition; with $n = 3, 4, 5, 6, 9,$ or 10, we obtain immediately 6 such numbers.

From this discussion, we see that perfect numbers are positive integers that are neither deficient nor abundant. Perhaps it was for this reason that Nicomachus observed that beautiful and extraordinary things are rare and easily tallied, while ugly and disorderly things proliferate like an infestation. So the deficient and abundant numbers are everywhere and irregular.

We now summarize the state so far of the various conjectures due to Nicomachus: the first and third conjecture were proven false; the fourth and fifth conjectures can be combined into a single conjecture (based on results from Euler); this and the second conjecture are still open and comprise the problem of perfect numbers.

2.6.   Weird Numbers and Semiperfect Numbers

We can observe the following distinction between different abundant numbers; some of the abundant numbers, such as for example 12, of course exceed their divisor sum, but nevertheless can be obtained as a sum of some of the divisors: $12 = 2 + 4 + 6 = 1 + 2 + 3 + 6$; other abundant numbers, such as 70 (with divisor sum $1 + 2 + 5 + 7 + 10 + 14 + 35 = 74$) do not have this property.

This distinction gives rise to a new definition: numbers of the former type (including as a special case the perfect numbers) are called *semiperfect numbers*, abundant numbers of the latter type are called *weird numbers*. The first ten semiperfect numbers are 6, 12, 18, 20, 24, 28, 30, 36, 40, and 42.

Every perfect number is semiperfect, and every multiple of a semiperfect number is also semiperfect; therefore there are infinitely many semiperfect numbers. Every number of the form $n = 2^k p$ with p an odd prime is semiperfect as long as $p < 2^{k+1}$. Numbers of the form $2^k(2^{k+1} - 1)$ are semiperfect; in particular, if $2^{k+1} - 1$ is a Mersenne prime, then such a number is perfect.

The smallest odd semiperfect number is 945, which was first observed by C. Friedman in 1993. In fact, $945 = 3^3 \times 5 \times 7$, $\frac{\sigma(945)}{2} - 945 = 15 = 1 + 5 + 9$.

A semiperfect number with no semiperfect factors is called a *primitive semiperfect number*. The first ten primitive semiperfect numbers are 6, 20, 28, 88, 104, 272, 304, 350, 368, and 464. There are infinitely many primitive semiperfect numbers; for example, $2^k p$ whenever p is an odd prime and $2^k < p < 2^{k+1}$ (there is always a prime number in this range by Bertrand's postulate).

The Hungarian mathematician Paul Erdős (1913–1996) was well-versed in number theory, and it was inevitable that he would pay some attention to the question of perfect numbers (Fig. 2.6). He proved that there are infinitely many odd primitive semiperfect numbers.

We turn now to the weird numbers; the first ten weird numbers are 70, 836, 4030, 7192, 7912, 9272, 10430, 10570, 10792, and 10990. There are infinitely many weird numbers, for example $70p$ whenever $p \geq 149$ is a prime number. But odd weird numbers are much more scarce; any odd weird number must be larger than 10^{21}. In 1976,

Figure 2.6.   Hungarian mathematician Paul Erdös

Sidney Kravitz found the odd weird number

$$2^{56}(2^{61} - 1) \cdot 153722867280912929 \approx 2 \cdot 10^{52}.$$

Finally, we consider Gaussian perfect numbers; in fact, many theorists have tried to define an analogue to the perfect numbers in the ring $\mathbb{Z}[i]$ of Gaussian integers. We discuss here just one such generalization. Let ν be a Gaussian integer, and let

$$N : \mathbb{Z}[i] \longrightarrow \mathbb{Z}^{\geq 0}$$

be the usual norm on this ring. If

$$\sum_{\delta \mid \nu} N(\delta) = 2N(\nu),$$

then we call ν a Gaussian perfect number; similarly we define the kth-order Gaussian perfect numbers (see Section 2.10 of this chapter) by the identity

$$\sum_{\delta \mid \nu} N(\delta) = (k+1)N(\nu).$$

This definition was proposed by Alessandro Rezende De Macedo and M. Riley Zeigler at the 2015 West Coast Number Theory Conference in Pacific Grove, California. They called such numbers *norm perfect numbers* and identified the Gaussian perfect number $9 + 3i$. Note that

$$9 + 3i = 3(1 + i)(2 - i)$$

with divisors $1, 3, 1 + i, 2 - i, 3 + 3i, 6 - 3i, 3 + i$. The relevant sum therefore is

$$N(1) + N(3) + N(1 + i) + N(2 - i)$$
$$+ N(3 + 3i) + N(6 - 3i) + N(3 + i)$$
$$= 1 + 9 + 2 + 5 + 18 + 45 + 10$$
$$= 90$$
$$= N(9 + 3i).$$

Since the definition of norm perfect numbers is at once beautiful and concise, and such numbers are defined as elements of the ring of Gaussian integers, we elect here to call them Gaussian perfect numbers. Naturally, there are open questions concerning such numbers: apart from $9 + 3i$ and its complex conjugate and associates, what other Gaussian perfect numbers are there? Are there any kth-order Gaussian perfect numbers? Are there any criteria for the determination of such numbers? And of course, are there finitely or infinitely many such numbers?

2.7.   Ore Numbers and the Harmonic Mean

In 1948, the Norwegian mathematician Øystein Ore defined the *harmonic divisor numbers*, also referred to as *Ore numbers* (Fig. 2.7). These numbers are positive integers n such that the harmonic mean of the divisors of n is an integer. If we introduce the symbols

$$\tau(n) = \sum_{d \mid n} 1$$

to represent the number of divisors of n and $H(n)$ to represent the harmonic mean of the divisors of n, this is

$$H(n) = \frac{\tau(n)}{\sum_{d \mid n} \frac{1}{d}} = \frac{n\tau(n)}{\sigma(n)} \in \mathbb{Z}.$$

Figure 2.7. Norwegian mathematician Øystein Ore

For example, the harmonic mean of the divisors of 6 is

$$H(6) = \frac{4}{1 + \frac{1}{2} + \frac{1}{3} + \frac{1}{6}} = 2,$$

so 6 is an Ore number. Similarly, the harmonic mean of the divisors of 140 is

$$H(140) = \frac{12}{1 + \frac{1}{2} + \frac{1}{4} + \frac{1}{5} + \frac{1}{7} + \frac{1}{10} + \frac{1}{14} + \frac{1}{20} + \frac{1}{28} + \frac{1}{35} + \frac{1}{70} + \frac{1}{140}}$$
$$= 5,$$

so 140 is also an Ore number.

The computation of $H(n)$ is simplified by the observation that both τ and σ are multiplicative functions, so also H is a multiplicative function. For example, if first we calculate

$$H(4) = \frac{3}{1 + \frac{1}{2} + \frac{1}{4}} = \frac{12}{7},$$

$$H(5) = \frac{2}{1 + \frac{1}{5}} = \frac{5}{3}, \quad \text{and}$$

$$H(7) = \frac{2}{1 + \frac{1}{7}} = \frac{7}{4},$$

we can determine immediately that

$$H(140) = H(4)H(5)H(7) = \frac{12}{7} \cdot \frac{5}{3} \cdot \frac{7}{4} = 5.$$

The first thirteen Ore numbers are 1, 6, 28, 140, 270, 496, 672, 1638, 2970, 6200, 8128, 8190, and 18600.

Ore proved that every perfect number is also an Ore number. To prove this, suppose n is a perfect number; then $\sigma(n) = 2n$, so

$$H(n) = \frac{n\tau(n)}{\sigma(n)} = \frac{\tau(n)}{2}.$$

It suffices therefore to show that $\tau(n)$ is even whenever n is a perfect number. In fact, $\tau(n)$ is even if and only if n is not a square number, since if n is not square, then the divisors of n can be put into pairs d and $\frac{n}{d}$ where d runs through the range $1 \leq d < \sqrt{n}$. Since perfect numbers are never square numbers (by the Euclid–Euler theorem and (2.3)), it follows that every perfect number is an Ore number.

Ore conjectured that there are no nontrivial (i.e. $n \neq 1$) odd harmonic divisor numbers; of course, this would imply if it were true also that there are no odd perfect numbers, so this is one avenue for an attempted proof of the latter claim. It has been proved that any odd Ore number must have at least three prime factors, and must be larger than 10^{24}.

Østein Ore studied at the University of Oslo, and received his Ph.D. in 1924. Afterwards, he continued his studies at Göttingen University, where he learned about the new methods in abstract algebra from Emmy Noether (1882–1935); finally, he moved to to America to become a professor at Yale University, where he remained until his retirement. His main areas of research were in ring theory and graph theory; his interest to number theory was more of a hobby. Ore was also deeply interested in the history of mathematics, and wrote popular biographies of his fellow Norwegion Niels Henrik Abel and the Italian mathematician Cardano. In 1936, Ore was invited to deliver a lecture as plenary speaker at the International Congress of Mathematicians in Oslo, and he was elected to both the American Academy of Arts and Sciences and the Oslo Academy of Science.

2.8.   Variations on Ore Numbers

In this section, we present a variation on the definition of Ore numbers; we consider instead the harmonic mean of the nontrivial factors

$$h(n) = \frac{\tau(n) - 1}{\sum_{d|n} \frac{1}{d} - 1}$$

$$= \frac{n\tau(n) - n}{\sigma(n) - n}.$$

When $h(n)$ is an integer, we shall refer to n as a *new Ore number* (Fig. 2.8). If p is a prime number, then $h(p) = p$, so every prime number is a new Ore number. If n is a perfect number, then $h(n) = \tau(n) - 1$, so also every perfect number is a new Ore number.

If $n = pq$ is the product of two distinct prime numbers, then we can calculate directly

$$h(n) = \frac{3pq}{1 + p + q}$$

and it is easy to see that if both p and $q = 2p - 1$ are prime, then $n = pq$ is a new Ore number with $h(n) = q$. This criterion gives the new Ore numbers

$$2 \times 3, 3 \times 5, 7 \times 13, 19 \times 37, 31 \times 61, 37 \times 73, 79 \times 157, 97 \times 193, \ldots;$$

note that the perfect number 6 is among them.

Similarly, if $n = pqr$ is a product of three distinct prime numbers, then

$$h(n) = \frac{7pqr}{1 + p + q + r + pq + qr + rp}$$

and we get a new Ore number $n = pqr$ whenever p, $q = 2p - 1$, and $r = 2q - 1 = 4p - 3$ are all primes, with $h(n) = r$. This gives another collection of new Ore numbers,

$$2 \times 3 \times 5, 19 \times 37 \times 73, 79 \times 157 \times 313, 439 \times 977 \times 1753,$$
$$661 \times 1321 \times 2641, \ldots.$$

The smallest new Ore number with four distinct prime factors is

$$2131 \times 4261 \times 8521 \times 17041,$$

a number with sixteen digits.

Abacist vs. Algorismist

From Gregor Reisch: Margarita Philosophica
Strassbourg 1504

Figure 2.8.   *Number Theory and Its History* by Øystein Ore

The arguments above generalize as follows: if p is a prime number such that $2^k p - 2^k + 1$ is prime for all $0 \leq k \leq \alpha$ (for some positive integer α), then

$$n = \prod_{k=0}^{\alpha} \left(2^k p - 2^k + 1\right) \qquad (2.5)$$

is a new Ore number, with $h(n) = 2^\alpha p - 2^\alpha + 1$. If n is a new Ore number so obtained, then we call n an *Ore prime of length* α.

Question 2.3. Are there infinitely many Ore primes?

Question 2.4. For fixed positive integer α, do there exist infinitely many Ore primes of length α?

Question 2.5. Evaluate the sum or series $\sum_{\substack{n \text{ is a new} \\ \text{Ore number}}} \frac{1}{n}$.

We also have the following propisition.

Proposition 2.1. *The only new Ore numbers of the form* $n = p^2 q$ *are* $28 = 2^2 \times 7$ *and* $117 = 3^2 \times 13$.

Proof.   If $n = p^2 q$, then

$$h(n) = \frac{5p^2 q}{1 + p + q + pq + p^2}.$$

We assume first that both $p, q \neq 5$. If $h(n)$ is an integer, it must be among the divisors of $5p^2 q$:

$$h(n) = 1, 5, p, q, pq, p^2, p^2 q, 5p, 5q, 5pq, 5p^2, \text{ or } 5p^2 q.$$

Evidently it cannot be any of the last six of these, since

$$\frac{1}{h(n)} \cdot \frac{5p^2 q}{1 + p + q + pq + p^2} < 1$$

for $h(n) = p^2 q, 5p, 5q, 5pq, 5p^2$, or $5p^2 q$.
  If $h(n) = pq$, then

$$5p = \frac{5p^2 q}{h(n)}$$

$$= 1 + p + q + pq + p^2$$

$$> (1 + p + q)p$$

$$> 5p,$$

which is a contradiction.

If $h(n) = p$, then

$$4pq = 1 + p + q + p^2.$$

Multiply both sides of this equation by 16 and rearrange to get the polynomial equation

$$(4p - 1)(16q - 4p - 5) = 21,$$

with no solutions.

If $h(n) = q$, then

$$5p^2 = \frac{5p^2 q}{h(n)}$$

$$= 1 + p + q + pq + p^2,$$

equivalently

$$(1 + p)(1 + q) = 4p^2,$$

which implies that both p^2 divides $1 + q$ and $1 + p \leq 4$; substituting $p = 2$ does not lead to an integer value of q, while $p = 3$ implies $q = 8$ is not prime.

With $h(n) = p^2$, we obtain the first solution: if

$$5q = \frac{5p^2 q}{h(n)} = 1 + p + q + pq + p^2,$$

it follows that $p = 2$ or 3; $p = 2$ does not produce an integer solution q, but if $p = 3$, then $q = 13$ is prime. This corresponds to $n = 117$.

Finally, suppose $h(n) = 5$ or 1. Then

$$1 + p + q + pq + p^2 = (1 + p)(1 + q) + p^2 = p^2 q \quad \text{or} \quad 5p^2 q.$$

This implies that p^2 divides $1 + q$, say $1 + q = Ap^2$. Then substituting in the above formula yields

$$(1 + p) \cdot Ap^2 + p^2 = p^2(Ap^2 - 1) \quad \text{or} \quad 5p^2(Ap^2 - 1),$$

and canceling the p^2 terms gives

$$(1 + p)A + 1 = Ap^2 - 1 \quad \text{or} \quad 5(Ap^2 - 1).$$

Therefore

$$1 \equiv -1 \quad \text{or} \quad -5 \pmod{A},$$

which means that A divides 2 if $h(n) = 5$ and A divides 6 if $h(n) = 1$. We check the cases $A = 1$ or 2 and $A = 1$, 2, 3, or 6 by inspection and find only a single new solution, $p = 2$, $q = 7$, $n = 28$ with $A = 2$ and $h(n) = 5$.

This completes the argument when both $p, q \neq 5$. And a similar analysis shows that there are no new solutions when $p = 5$ or $q = 5$, so we are done. $\qquad\qquad\qquad\qquad\qquad\qquad\qquad\qquad\qquad\qquad\square$

2.9.   Amicable Numbers

Amicable numbers are pairs of positive integers such that the sum of the positive divisors of one is equal to the other. Trivially, every perfect number forms an amicable number pair with itself; generally we require that the two integers be distinct. The notion of amicable numbers is often attributed to Pythagoras, who is said to have discovered the first amicable number pair 220 and 284 (Fig. 2.9):

$$220 = 1 + 2 + 4 + 71 + 142,$$

$$284 = 1 + 2 + 4 + 5 + 10 + 11 + 20 + 22 + 44 + 55 + 110.$$

Incidentally, the number 220 also makes an appearance in the Bible, in Genesis, where Jacob is said to have given 220 sheep to his brother Esau as a token of his love.

In later generations, the amicable numbers were considered to have a certain mystical significance, and played a role in various magical and astrological beliefs. During the European Middle Ages, many Arabic mathematicians made investigations into the amicable numbers. Notable among them, the Syrian Thābit ibn Qurra (ca. 826–901), a Syrian mathematician and translator who pioneered a method for the discovery of amicable numbers (Fig. 2.10).

Thābit ibn Qurra's Theorem. *If $n > 1$ such that all of*

$$p = 3 \times 2^{n-1} - 1,$$

$$q = 3 \times 2^n - 1, \quad \text{and}$$

$$r = 9 \times 2^{2n-1} - 1$$

are prime, then the numbers $2^n pq$ and $2^n r$ form a pair of amicable numbers.

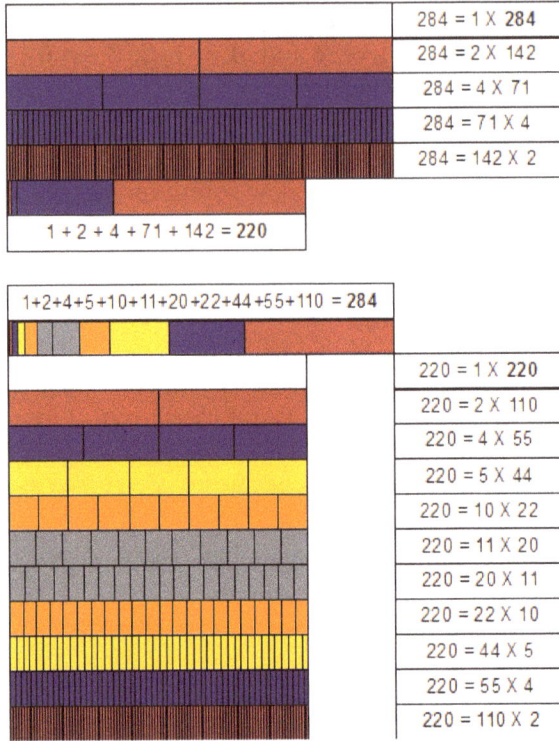

Figure 2.9. Depiction of the amicable pair (220, 284) with colored blocks

When $n = 2$, this corresponds to the amicable number pair discovered by the Pythagoreans. Neither Ibn Qurra nor any other Arabic mathematicians of the period were able to produce more amicable numbers using this method. In fact, there do exist such numbers, but they are large and require some effort to uncover. Many years later, Ibn Qurra's theorem was rediscovered by French mathematicians leading to the determination of new amicable numbers.

In 1636, more than 2000 years after the first amicable numbers were discovered, Fermat discovered that the pair of numbers 17296 and 18416 is also amicable; two years later, Descartes added another pair: 9363584 and 9437056. These two pairs correspond, respectively, to $n = 4$ and $n = 7$ in Ibn Qurra's theorem. Some sources also suggest that the amicable pair discovered by Descartes was in fact found much earlier by the Iranian mathematician Muhammad Baqir Yazdi in the 16th century, during the Safavid dynasty.

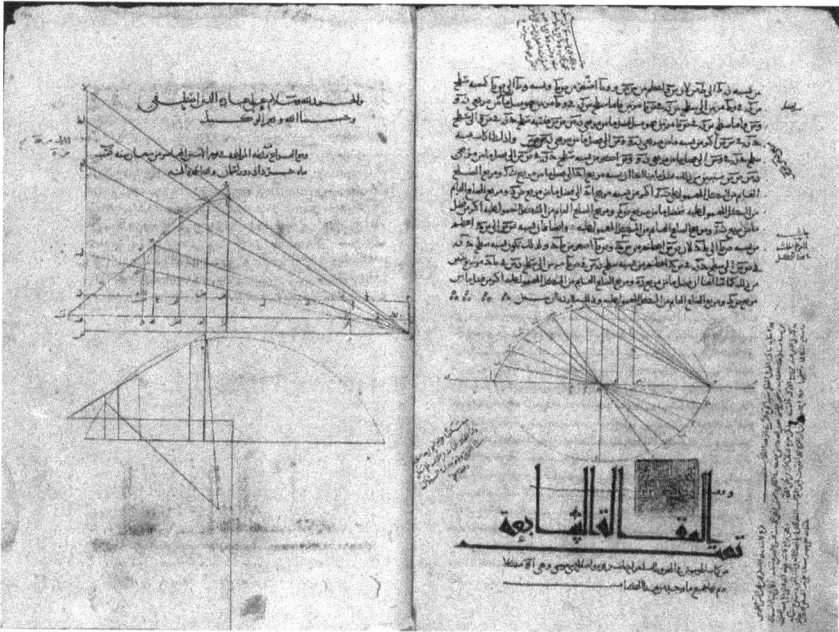

Figure 2.10. Translation by Thābit ibn Qurra of *Conic Sections*, by Apollonius

Naturally, Euler provided a generalization of Ibn Qurra's theorem, in 1747, and almost immediately found thirty new pairs of amicable numbers. Throughout his life, he found more that sixty such pairs; the smallest are 2620 and 2924,5020 and 5564, 6232 and 6368,10744 and 10856, and 12285 and 14595, all of them smaller than the pair that Fermat had already discovered. But in fact the second smallest pair of amicable numbers is 1184 and 1210, which was discovered in 1866 by an 18-year-old Italian named B. Nicolò I. Paganini. With modern computers, we can verify that the pair that Fermat found is the eighth smallest pair of amicable numbers.

The generalization due to Euler is as follows.

Euler's Rule. Suppose $n > m > 0$ are integers such that

$$p = (2^{n-m} + 1) \times 2^m - 1,$$
$$q = (2^{n-m} + 1) \times 2^n - 1,$$
$$r = (2^{n-m} + 1)^2 \times 2^{m+n} - 1$$

are all prime. Then the numbers $2^n pq$ and $2^n r$ are amicable. When $m = n - 1$, this is identical with Ibn Qurra's criterion.

Unfortunately, except for the cases $(m, n) = (1, 8)$ and $(29, 40)$, no additional amicable pairs have been generated by Euler's rule. By 1946, people had found 390 pairs of amicable numbers by hand. After the advent of the computer age, amicable numbers came in like a deluge. As of August 2017, more than 100 million amicable pairs have been discovered — $1,220,347,955$, to be exact. The following amicable pair comprising numbers with 36 digits and 42 digits, respectively, was found by Paul Bratley and John McKay in 1968:

$$a = 35380438442246018396504460782113062$$

$$b = 3538081696831696831682734954962738946069375.$$

In 1995, Paul Erdős (1913–1996) proved that the natural density of the set of all amicable numbers is zero. However, we still do not know whether or not there are infinitely many amicable pairs; moreover, every known amicable pair consists of two numbers of the same parity. Are there any amicable pairs containing to new numbers of opposite parity, or even relatively prime to one another?

Perhaps even more interesting is that people have been able to discover *sociable numbers*. A sequence of sociable numbers, or a *sociable chain of length n*, or a *sociable n-chain* is a collection of finitely many integers $n_1, \ldots, n_k$, such that

$$\sum_{\substack{d < n_1 \\ d | n_1}} d = n_2$$

$$\vdots$$

$$\sum_{\substack{d < n_j \\ d | n_j}} d = n_{j+1}$$

$$\vdots$$

$$\sum_{\substack{d < n_k \\ d | n_k}} d = n_1.$$

The definition of a sociable chain was introduced in 1918 by the self-taught Belgian mathematician Paul Poulet (1887–1946). He also presented two such chains; the first is

$$\cdots \longrightarrow 12494 \longrightarrow 14288 \longrightarrow 15472 \longrightarrow 14536 \longrightarrow 14264 \longrightarrow \cdots,$$

which remains to this day the only known sociable chain of length 5. The second has length 28:

$$\cdots \longrightarrow 14316 \longrightarrow 19116 \longrightarrow 31704 \longrightarrow 47616 \longrightarrow 83328 \longrightarrow 177792$$
$$\longrightarrow 295488 \longrightarrow 629072 \longrightarrow 589786 \longrightarrow 294896 \longrightarrow 358336 \longrightarrow 418904$$
$$\longrightarrow 366556 \longrightarrow 274924 \longrightarrow 275444 \longrightarrow 243760 \longrightarrow 376736 \longrightarrow 381028$$
$$\longrightarrow 285778 \longrightarrow 152990 \longrightarrow 122410 \longrightarrow 97946 \longrightarrow 48976 \longrightarrow 45946$$
$$\longrightarrow 22976 \longrightarrow 22744 \longrightarrow 19916 \longrightarrow 17716 \longrightarrow \cdots.$$

Even with the use of modern computers, this remains the longest known sociable chain. Poulet also discovered 43 new multiperfect numbers.

So far, a total of 1593 sociable chains have been discovered. Of these, 1581 are 4-chains; for example

$$\cdots \longrightarrow 1264460 \longrightarrow 1547860 \longrightarrow 1727636 \longrightarrow 1305184 \longrightarrow \cdots.$$

Also, five 6-chains, four 8-chains, one 9-chain, and one 28-chain. On the other hand, no one has ever discovered a 3-chain, or been able to prove that no such chains exist.

2.10. Multiply Perfect Numbers

As we have already seen, there is a bijective correspondence between the even perfect numbers and the Mersenne primes, and the search for Mersenne primes has become a venerable problem not only in mathematics, but also in computer science. The infinitude of such numbers seems at this time to be an immortal riddle, among the oldest and most difficult problems in all of mathematics. Meanwhile, many mathematicians have endeavored to find a suitable generalization of perfect numbers; for example, why not consider positive integers that only divide the sum of the proper divisors? In other

Figure 2.11.   The American mathematician Robert Carmichael

words, we add a coefficient to the right hand side of (1.1) to obtain

$$\sum_{\substack{d<n \\ d|n}} d = kn$$

or

$$\sigma(n) = (k+1)n,$$

for some positive integer k. Numbers satisfying this relation are called $(k+1)$-*multiply perfect numbers*, or sometimes $(k+1)$-*perfect numbers* or *multiply perfect numbers of order* $k+1$. When $k=1$, the 2-perfect numbers are simply the familiar perfect numbers. Various mathematicians have contributed to the search for $(k+1)$-multiply perfect numbers with $k>1$, including Fibonacci, Mersenne, Descartes, Fermat, and, later, Lehmer, Robert Carmichael (1879–1967), and Poulet (Fig. 2.11). Some of them were not able to find any, others found several; but in general, the results were scattered and haphazard, without the elegance of the correspondence between perfect numbers and Mersenne primes.

The first person to find a $(k+1)$-multiply perfect number with $k>1$ was the Welsh mathematician Robert Recorde (1512–1558) of Cambridge University, who discovered that 120 is a perfect number of order 3, as recorded in his book *The Whetstone of Witte* (1557)

Figure 2.12. The Welsh mathematician Robert Recorde

(Fig. 2.12). In the same book, he used the symbol "=" to indicate equality for the first time, and introduced the symbols "+" and "−" for addition and subtraction, respectively, to the English-speaking world. In 1631, Mersenne rediscovered this 3-perfect number, which is verified by hand by the calculation

$$1 + 2 + 3 + 4 + 5 + 6 + 8 + 10 + 12 + 15 + 20$$
$$+ \, 24 + 30 + 40 + 60 + 120 = 3 \times 120.$$

Mersenne also challenged Fermat to find a 3-perfect number other than 120. Six years later, Fermat completed this challenge and determined that also 672 is 3-perfect. That was in 1637, the same year that he wrote down Fermat's last theorem. Fermat also observed that if $m \geq 1$ such that $p = \frac{2^{m+3}-1}{2^m+1}$ is a prime number, then $n = 3 \times 2^{m+2}p$ is 3-perfect. The two 3-perfect numbers 120 and 672 correspond to $m = 1$, $p = 5$ and $m = 3$, $p = 7$, respectively, but so far no other 3-perfect numbers have been found using this criterion.

Instead, another French mathematician named André Jumeau (possibly referred to also as M. de Sainte Croix) used a different method to discover in the following year the third smallest 3-perfect number. $523776 = 2^9 \times 3 \times 11 \times 31$. He challenged Descartes to produce another, and Descartes found the 10 digit 3-perfect number 1476304896 in the same year. In 1639, Mersenne discovered a 9 digit 3-perfect number 459818240. Today, we recognize these as the fifth and fourth smallest 3-perfect numbers, respectively. In 1642, Fermat found the sixth smallest: 51001180160, with 11 digits. Descartes observed that Fermat's method could not work to find the third smallest 3-perfect number.

After that, no new 3-perfect numbers were found, and people came to believe that perhaps there are no others. On the other hand, it is very easy to prove that if m is an odd perfect number, then $n = 2m$ is 3-perfect: since σ is multiplicative, if m is odd and $\sigma(m) = 2m$, then

$$\sigma(n) = \sigma(2m) = \sigma(2)\sigma(m) = 3 \times 2m = 3n.$$

Since each of the six known 3-perfect numbers are multiples of 4, if it can be proved that there is no seventh 3-perfect number, this would prove also that there are no odd perfect numbers, resolving one half of the perfect number problem.

The French mathematicians of the 17th century also found multiply perfect numbers of other orders. Descartes for example discovered the six 4-perfect numbers 30240, 32760, 23569920, 142990848, 66433720320, and 403031236608, and the two 5-perfect numbers 14182439040 and 31998395520. Using modern computers, we can identify these as the first, second, fourth, sixth, ninth, and eleventh 4-perfect numbers, and the smallest two 5-perfect numbers. However, mathematicians were not able to determine any completely general rule for the determination of k-perfect numbers, perhaps because such numbers are very rare. We mention two open questions concerning the k-multiply perfect numbers.

Question 2.6. Are there infinitely many k-multiply perfect numbers?

Question 2.7. Are there any odd k-multiply perfect numbers?

2.11.   Three Further Generalizations

We turn now to another generalization of the perfect numbers; these are the superperfect numbers, first defined in 1969 by the Indian mathematician D. Suryanarayana. Suryanaryana taught at the School of Computer Science and Engineering at Vishnu Institute of Technology, but he was also fond of number theory. The *superperfect numbers* are positive integers n satisfying

$$\sigma^2(n) = \sigma(\sigma(n)) = 2n.$$

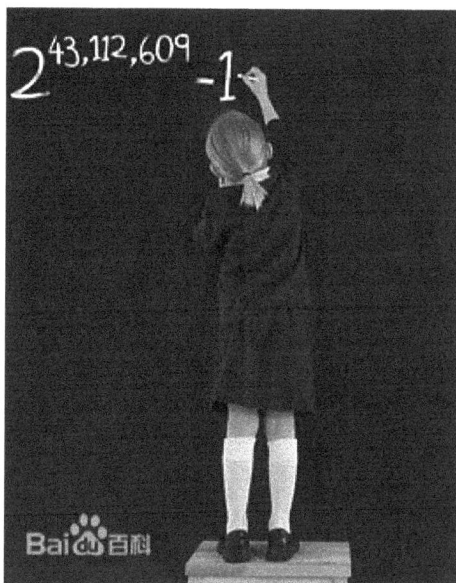

Figure 2.13. The 47th Mersenne prime

The first eight superperfect numbers are 2, 4, 16, 64, 4096, 65536, 262144, and 1073741824. For example,

$$\sigma^2(2) = \sigma(3) = 4 = 2 \times 2, \quad \text{and}$$
$$\sigma^2(16) = \sigma(31) = 32 = 2 \times 16.$$

If n is an even superperfect number, then n must be a power 2^k of 2 such that $2^{k-1} - 1$ is a Mersenne prime (Fig. 2.13). Are there any odd superperfect numbers? Nobody knows. Any odd superperfect numbers must be a square number n such that at least one of n or $\sigma(n)$ has at least three distinct prime factors; there are no odd superperfect numbers smaller than at least 7×10^{24}.

This definition can easily be generalized further to include the *m-superperfect numbers*; these are positive integers n satisfying

$$\sigma^m(n) = 2n.$$

The perfect numbers and superperfect numbers are the special cases $m = 1$ and $m = 2$ of m-superperfect numbers respectively; when $m \geq 3$, it can be shown that there do not exist any even m-superperfect numbers.

Finally, we have the most general definition in this direction. The (m, k)-*perfect numbers* are positive integers n satisfying

$$\sigma^m(n) = kn.$$

So the perfect numbers are $(1, 2)$-perfect numbers, the k-multiply perfect numbers are $(1, k)$-perfect numbers, the superperfect numbers are $(2, 2)$-perfect numbers, and the m-superperfect numbers are $(m, 2)$-perfect numbers. The first three $(2, 3)$-perfect numbers and $(2, 4)$-perfect numbers respectively are 8, 21, 512, and 15, 1023, 29127.

In a different direction of generalization, there are the k-*hyperperfect numbers*. These are positive integers n satisfying the equation

$$n = 1 + k \sum_{\substack{1 < d < n \\ d \mid n}} d,$$

or equivalently

$$n = 1 + k(\sigma(n) - n - 1).$$

The k-hyperperfect numbers were first introduced in a paper by Daniel Minoli and Robert Bear in 1975. The 1-hyperperfect numbers are simply the perfect numbers; the set of all k-hyperperfect numbers for all $k \geq 1$ is referred to generically as the set of hyperperfect numbers. The first seven hyperperfect numbers are 6, 21, 28, 301, 325, 496, and 697, with $k = 1, 2, 1, 6, 3, 1,$ and 12 respectively.

It is easy to prove by inspection that if $k > 1$ is odd and both

$$p = (3k + 1)/2, \quad \text{and}$$

$$q = 2p + 3 = 3k + 4$$

are both prime, then $n = p^2 q$ is k-perfect. Judson McCranie proposed the conjecture in 2000 that when $k > 1$ is odd, every k-hyperperfect number is of this form. In the same paper, McCranie identified by computer search every hyperperfect number up to 10^{11}.

It is also easy to verify that if p and q are distinct odd primes such that

$$k(p + q) = pq - 1$$

for some integer k, then $n = pq$ is k-hyperperfect.

Finally, suppose $k > 0$ and $p = k+1$ is prime. Then if $q_m = p^m - p + 1$ is prime for some $m > 1$, then $n = p^{m-1}q_m$ is k-hyperperfect. When $k = 1$, $p = 2$ this is Euclid's criterion. As another example, consider $k = 2$, $p = 3$. Then $m = 2$, 4, 5, and 6 generate the numbers $q_m = 7$, 79, 241, and 727 respectively, all of which are prime. This gives the 2-hyperperfect numbers 21, 2133, 19621, and 176661.

In the fall of 2020, the author of this book introduced two further generalizations of perfect numbers in the course of his regular graduate number theory seminary at Zhejiang University, in Hangzhou, China. We first consider positive integers n satisfying

$$n = \sum_{\substack{d < n \\ d \mid n}} d(1 + \mu(d)),$$

where $\mu : \mathbb{Z}^{>0} \longrightarrow \mathbb{Z}$ is the Möbius function

$$\mu(n) = \begin{cases} 0 & \text{if } n \text{ is not squarefree,} \\ (-1)^k & \text{if } n \text{ is squarefree and } n = p_1 \cdots p_k. \end{cases}$$

We call such integers n *Möbius perfect numbers* of the first kind. Peng Yang checked that among $n \leq 10^7$, the only Möbius perfect numbers of the first kind are 1, 42, and 460.

If

$$\sum_{\substack{d < n \\ d \mid n}} d(1 - \mu(d)) = n + 2,$$

then we call n a *Möbius perfect number of the second kind*.

It is easy to see that if both p and $2^p - 1$ are prime (i.e. $2^p - 1$ is a Mersenne prime), then $n = 2^p(2^p - 1)$ is a Möbius perfect number of the second kind; equivalently, if n is an even perfect number, then $2n$ is a Möbius perfect number of the second kind. For other n, Peng Yang also checked by computer that there are only two in the range $n \leq 10^7$. These are 765 and 1450. We proved that $1450 = 2 \times 5^2 \times 29$ is the only Möbius perfect number of the second kind of the form $n = 2p^2q$ with p, q distinct odd primes. Zhongyan Shen proved that if a positive integer n has exactly two prime factors, then it is a Möbius perfect number of the second kind if and only if it is twice a perfect number; we have not been able to discover whether or not there are any odd Möbius perfect number of the second kind other than 765.

2.12. $\mathcal{S}$-perfect Numbers

Also in the fall of 2020, the American graduate student of Zhejiang University Tyler Ross introduced in our seminar the $\mathcal{S}$-*perfect numbers*, and proved some results about them.

Let $\mathcal{S} \subset \mathbb{Z}$ be any collection of integers, and let $n \in \mathbb{Z}$ with $|n| > 1$. Label the positive divisors of n as

$$1 = d_0 < d_1 < \cdots < d_k < d_{k+1} = |n|.$$

Then we call n an $\mathcal{S}$-*perfect number of the first kind* if there exist integers $\lambda_1, \ldots, \lambda_k \in \mathcal{S}$ such that

$$1 + \sum_{j=1}^{k} \lambda_j d_j = n,$$

and an $\mathcal{S}$-*perfect number of the second kind* if there exist integers $\lambda_0, \ldots, \lambda_k \in \mathcal{S}$ such that

$$\lambda_0 + \sum_{j=1}^{k} \lambda_j d_j = n.$$

If n is an $\mathcal{S}$-perfect number, we refer to the sum $n = 1 + \sum_{j=1}^{k} \lambda_j d_j$ (or $n = \lambda_0 + \sum_{j=1}^{k} \lambda_j d_j$) as an $\mathcal{S}$-*presentation* of n, or simply a presentation of n when $\mathcal{S}$ is fixed.

The $\mathcal{S}$-perfect numbers generalize the perfect numbers ($\mathcal{S} = \{1\}$) and several other previous generalizations of the perfect numbers that we have already encountered. The $\{0, 1\}$-perfect numbers of the second kind are exactly the semiperfect numbers. When $k \geq 1$, the $\{k\}$-perfect numbers of the first kind are the k-hyperperfect numbers. The Möbius perfect numbers discussed in the previous section are special cases of $\{0, 1, 2\}$-perfect numbers.

In the following remarks, we consider only $\mathcal{S}$-perfect numbers of the first kind. Note first that for most integers $n > 1$ it is easy to find a set $\mathcal{S} \subset \mathbb{Z}$ such that n is $\mathcal{S}$-perfect.

Proposition 2.2. *If $n \in \mathbb{Z}(|n| > 1)$ has at least two prime factors, then there exists a finite set $\mathcal{S} \subset \mathbb{Z}$ with $\#\mathcal{S} \leq \tau(n) - 2$ such that n is $\mathcal{S}$-perfect. If $n = p^k$ for some $p \in \mathbb{P}$, $k \geq 1$, then n is not $\mathcal{S}$-perfect for any $\mathcal{S} \subset \mathbb{Z}$.*

Proof. If n has at least two prime factors, with proper divisors

$$1 = d_0 < d_1 < \cdots < d_M < |n|,$$

then $\gcd(d_1, \ldots, d_M) = 1$. It follows that the linear diophantine equation

$$\sum_{m=1}^{M} d_m x_m = n - 1$$

has solutions. As for the second statement, it is obvious that no prime or prime power can be $\mathcal{S}$-perfect for any $\mathcal{S} \subset \mathbb{Z}$. $\square$

We have also the following four theorems. The statements of the latter two theorems make use of the arithmetic function $\operatorname{ord}_2(n) = \max(k \geq 0 : 2^k \text{ divides } n)$.

Theorem 2.2. *If $m > 0$, then there exist infinitely many $\{0, m\}$-perfect numbers.*

Theorem 2.3. *Suppose $p_1, \ldots, p_t$ are distinct odd primes and $k_1, \ldots, k_t \geq 1$ are not all even. Then $n = 2^k p_1^{k_1} \cdots p_t^{k_t}$ is $\{1, -1\}$-perfect for all but finitely many $k \geq 1$. Conversely, if $n = 2^k p_1^{k_1} \cdots p_t^{k_t}$ is $\{1, -1\}$-perfect for some distinct odd primes $p_1, \ldots, p_t$ and integers $k \geq 0$, $k_1, \ldots, k_t \geq 1$, then at least one of $k_1, \ldots, k_t$ is odd.*

Theorem 2.4. *Fix $m \geq 1$ and set $\beta = \operatorname{ord}_2(m + 1)$. If $k, \alpha \geq 1$, $p \in \mathbb{P}^{odd}$, and both $2^k p$ and $2^{k+\alpha} p$ are $\{-1, m\}$-perfect, then $2^\alpha \equiv 1 \pmod{\frac{m+1}{2^\beta}}$. Conversely, if $k \geq \beta$, $2^\alpha \equiv 1 \pmod{\frac{m+1}{2^\beta}}$, and $2^k p$ is $\{-1, m\}$-perfect, then also $2^{k+\alpha} p$ is $\{-1, m\}$-perfect.*

Theorem 2.5. *Fix $m \geq 1$ and set $\beta = \operatorname{ord}_2(m + 1)$. Choose $\alpha \geq \beta$ such that $2^\alpha \equiv 1 \pmod{\frac{m+1}{2^\beta}}$. If p is a prime number satisfying $p \equiv 2(2^{\alpha+1} - 1) - 1 \pmod{2(m + 1)}$, then $2^k p$ is $\{-1, m\}$-perfect for infinitely many $k \geq 1$.*

We present here the proofs of the first and second of these theorems only; the proofs of the latter two theorems rely on similar arguments, but are more involved.

In fact, Theorem 2.2 is very easy to prove. Note that if $m > 0$ and n is $\{0, 1\}$-perfect with presentation $n = \sum$, then also $(m + 1)n$

is $\{0, 1\}$-perfect, with presentation

$$(m+1)n = \sum + mn.$$

Therefore it suffices to exhibit a single $\{0, m\}$-perfect number, for example:

$$n = (m+1)(m^2 + m + 1) = 1 + m(m+1) + m(m^2 + m + 1).$$

This completes the proof.

We turn next to the $\{-1, 1\}$-perfect numbers. The first few positive $\{-1, 1\}$-perfect numbers are 6, 12, 24, 28, 30, 40, 42, 48, 54, 56, 60, 66, 70, 78, 80, ... It is easy to see that every $\{-1, 1\}$-perfect number is either perfect or abundant; the smallest abundant number that is not also $\{-1, 1\}$-perfect is 18. The smallest odd abundant number is 945, which is also $\{-1, 1\}$-perfect, since

$$945 = 1 - 3 - 5 - 7 + 9 + 15 + 21 + 27 + 35 + 45$$
$$+ 63 + 105 + 135 + 189 + 315.$$

The following question remains open.

Question 2.8. Are there any odd abundant numbers that are not also $\{-1, 1\}$-perfect?

The proof of Theorem 2.3 requires a few lemmas.

Lemma 2.2. *If* $n \in \mathbb{Z}$, *then there exist integers* $k \geq 1$ *and* $\lambda_1, \ldots, \lambda_k \in \{1, -1\}$ *such that* $n = 1 + \sum_{j=1}^{k} \lambda_j 2^j$ *if and only if* $n \equiv 3 \pmod{4}$.

Proof. Obviously

$$1 + \sum_{j=1}^{k} \lambda_j \cdot 2^j \equiv 3 \pmod{4}$$

for all $k \geq 1$, $\lambda_1, \ldots, \lambda_k \in \{1, -1\}$. Suppose $n \equiv 3 \pmod{4}$, and choose $k \geq 1$ such that $3 - 2^{k+1} \leq n \leq 2^{k+1} - 1$. Then every $\Lambda =$

$(\lambda_1, \ldots, \lambda_k) \in \{-1, 1\}^k$ generates a distinct integer

$$N(\Lambda) = 1 + \sum_{j=1}^{k} \lambda_j \cdot 2^j \equiv 3 \pmod 4$$

in the range

$$3 - 2^{k+1} = 1 - \sum_{j=1}^{k} 2^j \leq N(\Lambda) \leq 1 + \sum_{j=1}^{k} 2^j = 2^{k+1} - 1.$$

Since there are exactly 2^k such integers, therefore $n = N(\Lambda)$ for some $\Lambda \in \{-1, 1\}^k$. $\qquad\square$

Lemma 2.3. *Suppose $n \in \mathbb{Z}$, and $p \in \mathbb{P}$ does not divide n. Then* (1) *if n is $\{-1, 1\}$-perfect, then so is np^k for all $k \geq 1$, and* (2) *if np is $\{-1, 1\}$-perfect, then so is np^{2k-1} for all $k \geq 1$.*

Proof. If (1) $n = \sum_1$ and $np^k = \sum_2$ $(k \geq 0)$ are $\{-1, 1\}$-presentations of n and np^k respectively, then $np^{k+1} = \sum_2 - np^k + p^{k+1} \sum_1$ is a presentation of np^{k+1}. Similarly, if (2) $np = \sum_1$ and $np^k = \sum_2$ $(k \geq 1)$ are presentations of np and np^k respectively, then $np^{k+2} = \sum_2 - np^k + p^{k+1} \sum_1$ is a presentation of np^{k+2}. $\qquad\square$

Lemma 2.4. *If $n \in \mathbb{Z}^{>k}$ is $\{1, -1\}$-perfect, then so is $2n$.*

Proof. If n is odd, this follows from Lemma 2.3. Suppose n is even and $n = 1 + \sum_{j=1}^{k} \lambda_j d_j$ is a presentation of n. Then $2n = 1 + \sum_{j=1}^{k} \lambda_j d_j + n$. The proper divisors of $2n$ missing from this sum all have the form $2d_j$ for some divisor d_j of n with $1 < d_j < n$ (here we require that n is even). Replace all such $\lambda_j d_j$ in the sum with $-\lambda_j d_j + \lambda_j(2d_j)$ to obtain a presentation of $2n$. $\qquad\square$

Proof of Theorem 2.3. In light of Lemmas 2.3 and 2.4, it suffices to show that $n = 2^k p$ is $\{1, -1\}$-perfect for all $p \in \mathbb{P}^{\mathrm{odd}}$ for some

$k \geq 1$. Choose (Lemma 2.2) $k_0 \geq 1$, $\lambda_1, \ldots, \lambda_{k_0} \in \{1, -1\}$ such that

$$1 + \sum_{j=1}^{k_0} \lambda_j 2^j = \begin{cases} p & \text{if } p \equiv 3 \pmod 4, \\ 3p & \text{if } p \equiv 1 \pmod 4. \end{cases}$$

Then

$$2^{k_0} p = 1 + \sum_{j=1}^{k_0} \lambda_j 2^j + (-1)^{(p+1)/2} p + \sum_{j=1}^{k_0-1} 2^j p$$

is a presentation of $2^{k_0} p$, as required.

Conversely, note that if $n > 1$ is $\{1, -1\}$-perfect with presentation

$$n = 1 + \sum_{j=1}^{k} \lambda_j d_j,$$

then

$$\sigma(n) = \sum_{j=1}^{k} (1 - \lambda_j) d_j + 2n$$

is even, since every $1 - \lambda_j = 0$ or 2. But it is well known that $\sigma(n)$ is even if and only if n is not square or twice a square. This completes the proof. □

Conjecture 2.1. *The set of positive $\{-1, 1\}$-perfect numbers has a natural density, identical to the density of the abundant numbers.*

Question 2.10. For what $m > 1$ (if any) are there infinitely many $\{1.m\}$-perfect numbers, not counting the classical perfect numbers?

2.13. The Golden Ratio Conjecture

On September 30th, 2017, the author posted the following remarks on Sina Weibo:

> A small discovery: the Parthenon in Athens is the model of classical beauty; its east and west sides have height and width nineteen meters and thirty-one meters respectively. The ratio

of these two numbers is about 0.613..., not so far off from the golden ratio 0.618.... Recently I happened to notice something about the perfect numbers, which were first introduced by the Pythagorean school. After 2500 years, we have discovered 49 perfect numbers; thirty of these terminate in the digit 6 and nineteen of them in the digit 8. I personally predict that also the fiftieth perfect number, when it is discovered, will be found to end with the digit 6. Moreover, if it turns out that there are infinitely many perfect numbers, perhaps the ratio between the number of them that terminate in 8 and the number of them that terminate in 6 will tend to the golden ratio.

A few days ago, I remembered these remarks with interest and went again to look at the table of even perfect numbers and tally up the number of them that terminate in 6 and 8, respectively. It felt as familiar as an old friend. Then I checked again the length and width of the sides of the Parthenon (Fig. 2.14). Unexpected discoveries and conjectures such as this one can come only from data. In fact, we have already seen in the proof of Property 4 in Section 2.1, that when $p = 2$ or $p \equiv 1 \pmod 4$, then the corresponding perfect number

Figure 2.14.   The Parthenon in Athens

terminates in 6; when $p \equiv 3 \pmod{4}$, then the corresponding perfect number terminates in 8.

At that time, I believed that the fiftieth Mersenne prime and corresponding perfect number would be discovered within five years. In fact, it took only three months: on December 27th, 2017, it was announced that a new Mersenne prime number had been found, given by $p = 77232917$. And indeed, the corresponding fiftieth even perfect number really does end in 6. In light of this the author puts forward here the following rather bold conjecture.

Conjecture 2.2. *There are infinitely many perfect numbers, and the ratio between the number of them that terminate in 8 and the number of them that terminate in 6 tends to the golden ratio.*

Obviously, the claim here that there infinitely many perfect numbers is linked with the irrational number called the golden ratio. It is interesting that both the concept of perfect numbers and the golden ratio most likely originated with Pythagoras or his follows, but they likely did not imagine that there was some connection between the two.

A year later, the 51st perfect number was discovered. It too terminates in 6. If we leave out the perfect number 6, which is the only perfect number generated by an even prime number, the number of perfect numbers generated by primes $p \equiv 1 \pmod{4}$ and $p \equiv 3 \pmod{4}$ are still 19 and 31.

We can express this conjecture as formula as follows:

$$\sum_{\substack{p \leq x \\ p \equiv 1 \ (\mathrm{mod}\,4) \\ 2^p - 1 \text{ is prime}}} 1 \approx 0.618 \cdots \times \sum_{\substack{p \leq x \\ p \equiv 3 \ (\mathrm{mod}\,4) \\ 2^p - 1 \text{ is prime}}} 1.$$

An explanation for the phenomenon discussed here is that the distribution of prime numbers among arithmetic progressions with the same difference is not uniform when we restrict the variable to prime values. For example, if we consider prime numbers of the form $4p + 1$ and $4p + 3$ as p varies across the prime numbers, computer analysis supports the approximate identity

$$\sum_{\substack{p \leq x \\ 4p+1 \text{ is prime}}} 1 \approx \frac{1}{2} \sum_{\substack{p \leq x \\ 4p+3 \text{ is prime}}} 1. \tag{2.6}$$

Of course, according to Dirichlet's theorem on primes in arithmetic progressions, the number of primes in arithmetic progressions with the same difference should in general be the same; that is, if $(k, m) = 1$, $1 \leq k < m$, then

$$\sum_{\substack{n \leq x \\ k+nm \text{ is prime}}} 1 \approx \frac{x}{\phi(m) \cdot \log(x)}$$

regardless of the choice of k. But this does not hold when n is restricted to the primes. Instead, we have the following conjectures, in addition to (2.6):

$$\sum_{\substack{p \leq x \\ p+2 \text{ is prime}}} 1 \approx \sum_{\substack{p \leq x \\ p+4 \text{ is prime}}} 1 \approx \frac{1}{2} \sum_{\substack{p \leq x \\ p+6 \text{ is prime}}} 1,$$

$$\sum_{\substack{p \leq x \\ 6p+1 \text{ is prime}}} 1 \approx \frac{3}{4} \sum_{\substack{p \leq x \\ 6p+5 \text{ is prime}}} 1,$$

$$\sum_{\substack{p \leq x \\ 8p+1 \text{ is prime}}} 1 \approx \frac{1}{2} \sum_{\substack{p \leq x \\ 8p+3 \text{ is prime}}} 1 \approx \frac{2}{3} \sum_{\substack{p \leq x \\ 8p+5 \text{ is prime}}} 1 \approx \frac{5}{6} \sum_{\substack{p \leq x \\ 8p+7 \text{ is prime}}} 1.$$

There is also the famous special case of the first Hardy–Littlewood conjecture, which says that

$$\sum_{\substack{p \leq x \\ p+2 \text{ is prime}}} 1 \approx 2C \frac{x}{(\log x)^2},$$

where

$$C = \prod_{p>2} \frac{p(p-2)}{(p-1)^2} \approx 0.660161.$$

In 1962, Paul Bateman and Roger Horn proposed a more general conjecture. Suppose $f_1, \ldots, f_m$ is a system of distinct irreducible polynomials with integer coefficients, and set $f = \prod_{1 \leq k \leq m} f_k$. Then the Bateman–Horn conjecture is

$$\sum_{\substack{n \le x \\ \text{every } f_k(n) \text{ is prime}}} 1 \approx \frac{C}{D} \cdot \frac{x}{(\log x)^{m+1}}. \tag{2.7}$$

The constants in this approximation are

$$D = \prod_{1 \le k \le m} \deg f_k$$

and

$$C = \prod_{p \in \mathbb{P}} \frac{1 - \frac{N(p)}{p}}{(1 - \frac{1}{p})^m}$$

where $N(p)$ is the number of solutions to $f(n) \equiv 0 \pmod{p}$. It can be shown that $N(p) < p$ for every prime p, from which it follows (with some work) that $C > 0$.

From (2.7) we can conclude that if if $(k, m) = 1$, $1 \le k < m$, then

$$\sum_{\substack{p \le x \\ k+mp \text{ is prime}}} 1 \approx \prod_{\substack{p|k \\ p>2}} \frac{p-2}{p-1} \sum_{\substack{p \le x \\ k+mp \text{ is prime}}} 1.$$

From this and (2.7), we get

$$\sum_{\substack{p \le x \\ k+mp \text{ is prime}}} 1 \approx 2 \prod_{\substack{p|km \\ p>2}} \frac{p-2}{p-1} \prod_{p>2} \frac{p(p-2)x}{(p-1)^2 (\log x)^{k+1}}.$$

Chapter 3

The Fibonacci Sequence

> We can conclude that all the
> knowledge we have of
> mathematics outside of Greece
> is due to the appearance of
> Fibonacci.
>
> *Gerolamo Cardano*

3.1. Fibonacci, Leonardo of Pisa

In the second half of the 12th century there appeared for the first time in Europe since the descent into the darkness of the Middle Ages an important and capable mathematician: Fibonacci (ca. 1170–1250). His real name was Leonardo Pisano, or Leonardo of Pisa. Pisa today is a city in Italy, but at that time there was not yet any such thing as Italy, and Pisa was a republic (Figs. 3.1 and 3.2).

Pisa is known around the world today especially for its leaning tower and white cathedral. Folklore has it that Galileo Galilei (1564–1642) performed experiments on falling bodies from atop the leaning tower. Whether or not it is true, this story has contributed to the flourishing of the tourism industry in Pisa. During the Middle Ages, Pisa was a strong maritime and military power, with a developed economy, and a center of trade. Along with Genoa, Amalfi (not far from Naples) on the western coast of Italy, and Venice on the shores

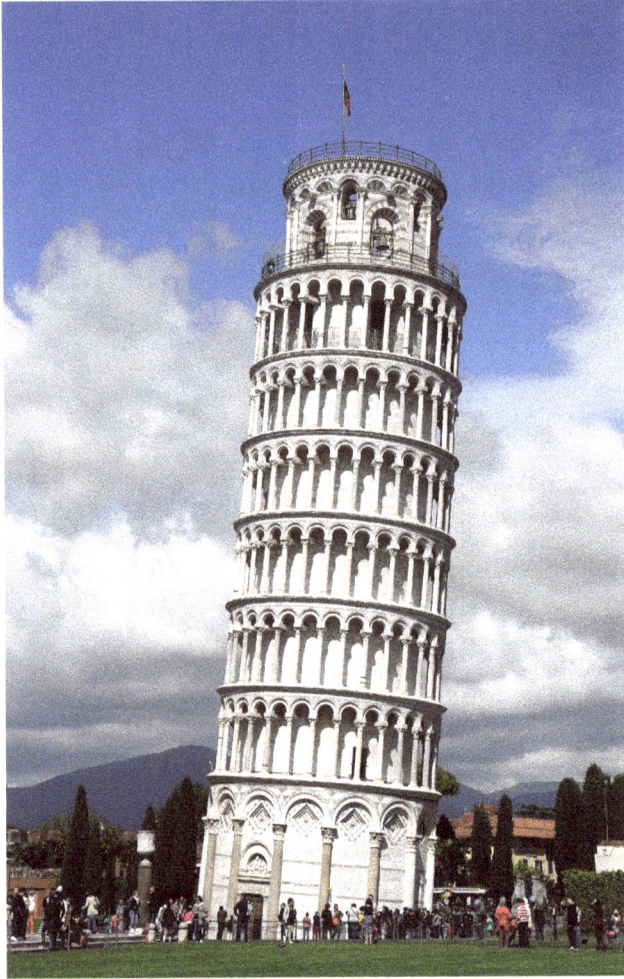

Figure 3.1.   The Leaning Tower of Pisa

of the Adriatic Sea, it was one of the four major maritime republics on the Apennine peninsula.

When Leonardo was born, the magnificent Cathedral of Pisa had already been built; construction of the leaning tower begin in 1173, when he was three years old. But it took two centuries to complete, so he did not live to see it finished. His father was a civil servant in the Republic of Pisa named Guglielmo Bonacci. Today he is almost always remembered as Fibonacci, which is an abbreviation of *Filius*

Figure 3.2. Portrait of Fibonacci

Bonacci, meaning *Son of Bonacci*. But he was first referred to by this name only in 1838 by the Italian mathematician Guglielmo Libri Carucci dalla Sommaja (1803–1869), also a notorious book thief who was responsible for the theft of about 30,000 books and manuscripts during his time as Inspector of Libraries in France.

Fibonacci followed his father in his travels to various destinations along the Mediterannean coast, including the port city of Bugia (now Béjaïa, in Algeria), where Bonacci served as consul to the local community of Pisan merchants. Fibonacci came into contact with the works of the ancient Greek mathematician Diophantus and the Persian mathematician Muhammad ibn Musa al-Khwarizmi; he also learned about the Hindu-Arabic system of numerals.

At the end of the 12th century, Fibonacci returned to Pisa, and remained there for the next quarter of a century. He spent his time in writing books, in which he introduced the Hindu-Arabic numerals and computation methods to Europe (a key moment in the history of global mathematics), discussed various geometric and algebraic problems and their commercial applications. His most significant mathematical achievements were in indeterminate analysis and number theory, in which disciplines he far surpassed any earlier or contemporaneous European mathematicians.

In around 1225, Fibonacci was invited by the Holy Roman Emperor Frederick II to become a court mathematician. As an aside, the life of Frederick II is the stuff of legends. The Italian poet

Giovanni Boccaccio included in his *De Mulieribus Claris* (*On Famous Women*) the biography of his biological mother, Constance, princess of Sicily. According to Boccaccio, it had been prophesied that her marriage would bring about the destruction of Sicily, and she spent the first thirty years of her life confined to a convent by the order of her father, until at last she became engaged to Henry, King of the Romans, whom she eventually married. She gave birth to Frederick, her first and only child, at the age of forty. She happened to be passing by a small town as she went in to labor, and ordered her entourage to set up a tent so that the women of the town could come and bear witness to the heritage of her child.

During his reign, Frederick vigorously developed the industry and commerce of Sicily, conquered northern Italy by force, and acquired control of Jerusalem in the course of the Sixth Crusade. He somehow also found time to master six languages (German, French, Latin, Greek, Hebrew, and Arabic), establish the University of Naples in 1224 as well as the Sicilian school of poetry, and compose a treatise on falconry entitled *The Art of Hunting with Birds*, and a collection of erotic poems. He was also a passionate supporter of the natural sciences; perhaps it was for the reason he felt the need to have mathematicians in his court. However, Fibonacci seems to have completed already most of his important mathematical work during his time in Pisa.

In his famous *Liber Abaci* (1202), Fibonacci mentions two problems that were already known in China as Zhang Qiujian's *Hundred Fowl Problem* and a special example of Qin Jiushao's theorem (the Chinese Remainder Theorem). On the other hand, the abacus of the title is not at all the same as the Chinese abacus but rather refers to a sand table used for calculating. Fibonacci uses for the first time in *Liber Abaci* a horizontal bar in fractions, as we still do today. More interesting is the so-called rabbit problem, which defines the famous Fibonacci sequence. It is, however, likely that the sequence was known much earlier to ancient Indian mathematicians.

In addition to *Liber Abaci*, Fibonacci also published books with the titles *Practical Geometry*, *Flowers*, *Mathematical Letter to Master Theodorus*,[a] and *The Book of Squares*. The last of these is the

---

[a]There was a Greek historian in the sixth century named Theodorus; it is not known, however, who was the Theodorus for whom this letter was written.

most original and interesting. It is concerned with indefinite equations, mainly the solutions in integers and rational numbers to the quadratic equations

$$x^2 \pm 5 = y^2.$$

For example, he found the solution $x = \frac{41}{12}$, $y = \frac{31}{12}$ to the equation $x^2 - 5 = y^2$.

Fibonacci also proved that $x^2 + y^2$ and $x^2 - y^2$ cannot simultaneously be square numbers, which means that the equation

$$x^4 - y^4 = z^2,$$

has no nontrivial (i.e. other than $x = y = z = 0$) integer solutions. Otherwise, if $x^4 - y^4 = z^2$ had some nontrivial integer solution, then also it must admit a solution with x and y relatively prime nonzero integers, from which it follows that $x^2 + y^2$ and $x^2 - y^2$ are both square.

The *Mathematical Letter to Master Theodorus* discusses again the hundred fowl problem, and introduces an interesting geometrical problem: to inscribe a regular pentagon in an isosceles triangle. Even from the modern perspective, this problem is not elementary. Fibonacci gives the construction through the solution of quadratic equations, an early example of algebraic methods applied to geometry. At the end of the letter he also provides a solution for a specific quintic equation.

Some time earlier, the Persian mathematician Omar Khayyam (1048–1122) had applied geometric methods to the solution of algebraic problems. Specifically, he decomposed a class of cubic equations into a combination of the quadratic equations for a circle and a parabola.

It is necessary to remark that symbolic algebra had not yet been invented at that time, so Fibonacci presented his results entirely in geometric terms. For example, he expresses a certain indefinite equation in *The Book of Squares* as follows:

> *if you subtract the area of a square from its four sides, what remains is three lengths.*

The number system in this and his other works is in base 60, reflecting the Arabic influence.

Due to the publication of these books and the popularization of the Fibonacci sequence, Fibonacci became the most influential mathematician between Diophantus in ancient Greek times and Fermat in the 17th century. The 16th century mathematician Gerolamo Cardano (1501–1576) remarked, "We can conclude that all the knowledge we have of mathematics outside of Greece is due to the appearance of Fibonacci."

3.2.   The Rabbit Problem

Let us now have a look at the famous rabbit problem (Fig. 3.3). Fibonacci stated this problem in his *Liber Abaci* as follows:

> A newly born pair of rabbits of opposite sexes is placed in an enclosure at the beginning of a year. After one month, they mate, and beginning with the second month, the female gives birth to a pair of rabbits of opposite sexes every month. Each new pair also gives birth to a pair of rabbits each month, starting from their second month following their birth. Find the number of pairs of rabbits in the enclosure after one year.

Obviously there is only a single pair of rabbits in both the first and second months; in the third month there are two pairs, one new and one old; in the fourth month there are three pairs, two old

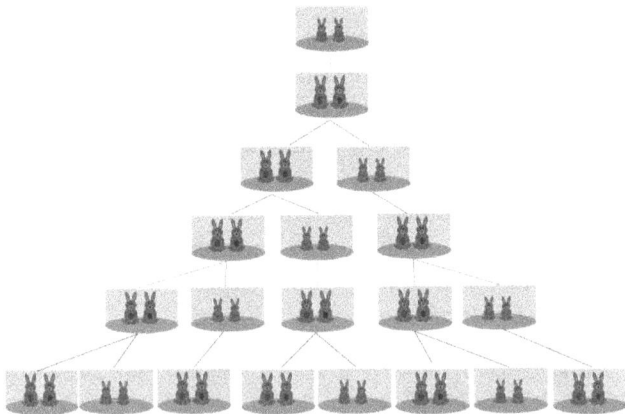

Figure 3.3.   The Rabbit Story

pairs and one new pair; in the fifth month there are five pairs, three old and two new; and so on.

Fibonacci could certainly not have imagined that his little rabbit problem would still attract the interest and attention of mathematicians around the world more than 800 years later. If we consider that the Chinese Remainder Theorem introduced by the Chinese mathematician Qin Jiushao (1202–1261) around more or less the same time was completely resolved both theoretically and computationally by its author, then the enduring depth of this innocuous seeming problem is all the more impressive.

In any case, it is easy to calculate the first twelve terms of the rabbit sequence:

$$1, 1, 2, 3, 5, 8, 13, 21, 34, 55, 89, 144.$$

If we let F_n denote the number of rabbit pairs in the nth month, this sequence is called the *Fibonacci sequence*, and its nth term F_n is called the nth *Fibonacci number*. The Fibonacci sequence is determined by

$$\begin{cases} F_0 = 0, \\ F_1 = 1, \\ F_n = F_{n-1} + F_{n-2} \quad \text{for } (n \geq 2). \end{cases}$$

It is necessary here to point out that as early as the second century BCE, the Sanskrit epics of India already contain instances of the Fibonacci numbers. In these epics, the long syllable is composed of two units of time, the short syllable a single unit. Then the question is, given a certain length unit of time, calculate the number of possible combinations of these two syllables (a special case of the composition problem). In other words, find the number of different ways to write a given integer as a sum using only the numbers 1 or 2 as summands and counting different orderings as different sums. For example, when $n = 5$ we have the sums

$$5 = 1+1+1+1+1 = 1+1+1+2 = 1+1+2+1$$
$$= 1+2+1+1 = 2+1+1+1 = 2+2+1$$
$$= 2+1+2 = 1+2+2,$$

in total eight possibilities corresponding to the Fibonacci number $F_6 = 8$. More generally, the Fibonacci F_{n+1} represents also the number ways to write an integer n as a composition of 1 and 2.

Yet another way to state this problem is the following: suppose you would like to climb a ladder with n rungs; how many ways are there to do this if you can freely ascend either one or two rungs at each step? And this analogy also provides the proof of the equivalence between this problem and the rabbit problem. Let a_n represent the number of different ways to ascend a ladder with n rungs. It is obvious that $a_1 = 1$, $a_2 = 2$. Consider $n \geq 3$; if we move up a single rung at the first step, then there are a_{n-1} ways to climb the remaining rungs. On the other hand, if we move up two rungs in the first step, then there remain a_{n-2} ways to climb the remaining rungs. We conclude that $a_n = a_{n-1} + a_{n-2}$, in agreement with the definition of the Fibonacci sequence. In other words, $a_n = F_{n+1}$.

The first explicit statement of the Fibonacci sequence in India was given by the prosodist and mathematician Virahanka, who lived and worked sometime between the sixth and eighth centuries. His work has not survived, but it was the subject of a commentary by Gopala sometime around the year 1135, before Fibonacci was born. Gopala clearly states the recurrence rule $F_{n+2} = F_{n+1} + F_n$ and enumerates this sequence up to its eighth term.

In later history, as scientific and technological knowledge advanced, people discovered that the Fibonacci sequence has direct applications in physics, chemistry, the theory of quasicrystals, and other disciplines. For this reason, the Fibonacci Association was established in 1963 by two California mathematicians. In the same year, the scientific journal *The Fibonacci Quarterly* began publication, dedicated to research results in this area. Since 1984, the Fibonacci Association has organized all around the world the biannual International Conference on Fibonacci Numbers and their Applications.

It is not uncommon to find the Fibonacci sequence in nature in unexpected places. As an example from the plant kingdom, many flowers tend to have a Fibonacci number of petals: plum bossoms have 5, delphiniums 8, marigolds 13, and asters 21; notably, daisies have either 34 or 55 or 89 petals. The Fibonacci numbers also turn up in the growth patterns of fingerprints, the seeds of a sunflower, the wings of a dragonfly.

In biology, there is Ludwig's law, which links the Fibonacci numbers to the growth of new shoots in branching plants. New shoots require a period of rest before the germination of a new bud; this period can be one year or several years. In the former case only, then the number of branches in a given year theoretically should be a Fibonacci number. There is also an interesting analogue to this law for the behavior of the stock market.

The rabbit sequence has been known as the Fibonacci sequence only since the second half of the 19th century, when Lucas gave it this name. The recurrence relation in their definition can be extended leftwards to negative indices to produce the sequence

$$\ldots, 5, -3, 2, -1, 1, 0, 1, 1, 2, 3, 5, \ldots$$

and in general

$$F_{-n} = (-1)^{n-1} F_n.$$

In the Pascal triangle (贾宪-杨辉三角 in China, after the mathematicians Yang Hui and Jia Xian), if you take the sum along parallel diagonals as in the figure, you obtain a Fibonacci number (Fig. 3.4). This formula is due to Lucas (1876):

$$F_{n+1} = \sum_{k=0}^{\lfloor n/2 \rfloor} \binom{n-k}{k}. \tag{3.1}$$

This can be proved by induction and the recurrence formula

$$\binom{n+1}{k+1} = \binom{n}{k} + \binom{n}{k+1},$$

for binomial coefficients. We give here instead a combinatorial proof. It is well known $\binom{n}{n-k}$ represents the number of paths from the origin

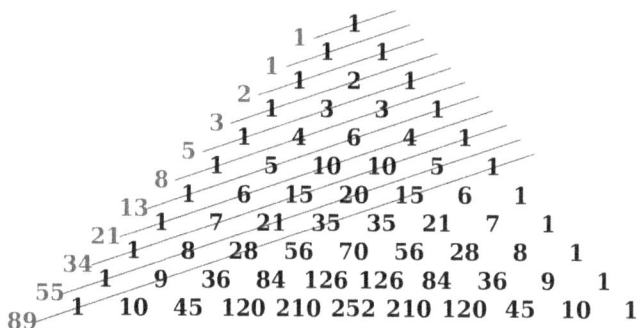

Figure 3.4.   Pascal's Triangle and Fibonacci Numbers

$(0,0)$ to the point $(k, n - k)$ using only the unit steps

$$(x, y) \longmapsto (x + 1, y), \text{ or}$$
$$(x, y) \longmapsto (x, y + 1),$$

since the total length of such a path contains n steps, exactly any $n - k$ of which must be the step $(x, y) \longmapsto (x, y + 1)$. The following argument is similar.

Proof of (3.1). Now consider instead the number of paths from the origin $(0,0)$ to the point $(0, n)$ using only the steps

$$(x, y) \longmapsto (x + 1, y), \text{ or}$$
$$(x, y) \longmapsto (x, y + 2),$$

This is identical to the ladder problem that we have already discussed, with solution F_{n+1}. But we also count the paths another way. There are $\binom{n-k}{k}$ different ways to construct the path using k steps of the form $(x, y) \longmapsto (x, y + 2)$ for each $0 \le k \le \lfloor n/2 \rfloor$. Add these together to get a total of

$$\sum_{k=0}^{\lfloor n/2 \rfloor} \binom{n - k}{k}$$

paths, so

$$F_{n+1} = \sum_{k=0}^{\lfloor n/2 \rfloor} \binom{n - k}{k}.$$

$\square$

It is worth pointing out, that if we take the diagonals along the Pascal triangle more obliquely we can obtain similar sums, for example

$$\sum_{k=0}^{\lfloor n/3 \rfloor} \binom{n - 2k}{k}.$$

The first several values of this sum as n takes nonnegative integer values are

$$0, 1, 1, 1, 2, 3, 4, 6, 9, 13, 19, 28, 41, 60, 88, \ldots,$$

which is identical to the *Narayana cow sequence* after that 14th century Indian mathematician Narayana Pandita. If we use G_n to indicate the nth number of this sequence, then it is obtained by the recurrence

$$G_0 = 0,$$
$$G_1 = G_2 = 1,$$
$$G_n = G_{n-1} + G_{n-3} \text{ for } (n \geq 3),$$

and we have

$$G_{n+1} = \sum_{k=0}^{\lfloor n/3 \rfloor} \binom{n - 2k}{k}.$$

The numbers G_{n+1} also represent the numbers presentations of n as a composition of 1 and 3. For example, if $n = 5$,

$$5 = 1+1+1+1+1 = 1+1+3 = 1+3+1 = 3+1+1,$$

corresponding to $G_6 = 4$. Also, G_{n+2} represents the number of presentations of n as compositions of 1 and 2 where there never occurs more than one 2 consecutively. For example, $G_7 = 6$ counts the compositions

$$5 = 1+1+1+1+1 = 1+1+1+2 = 1+1+2+1$$
$$= 1+2+1+1 = 2+1+1+1 = 2+1+2.$$

On the other hand, the extension of G_n to negative indices does not seem to behave as regularly as the similar extension of F_n. For $1 \leq n \leq 10$, the corresponding numbers G_{-n} are

$$\ldots, -2, 3, 0, -2, 1, 1, -1, 0, 1, 0.$$

3.3. General Terms and Limits

The Fibonacci sequence has many interesting properties; here we present a selection of identities and other results.

First, it is not hard to prove by induction on n or m the addition formula

$$F_{m+n} = F_m F_{n+1} + F_{m-1} F_n$$
$$= F_{m+1} F_n + F_m F_{n-1}, \tag{3.2}$$

from which it follows immediately that if d divides n, then F_d divides F_n. Therefore, F_n is composite whenever n is composite for all $n > 4$ (since $F_2 = 1$ and $F_4 = 3$).

The converse, however, is false: for example, with $p = 19$ prime, $F_{19} = 4181 = 37 \times 113$; and with $p = 53$, $F_{53} = 953 \times 55945741$.

In 1718, the French mathematician Abraham de Moivre (1667–1754), who spent almost his entire life in England, discovered the following remarkable explicit expression for F_n ($n \geq 1$) (Fig. 3.5):

$$F_n = \frac{1}{\sqrt{5}} \left(\left(\frac{1 + \sqrt{5}}{2} \right)^n - \left(\frac{1 - \sqrt{5}}{2} \right)^n \right).$$

This formula is generally known today as Binet's formula, after a later mathematician. Anyway, de Moivre was also the author of

Figure 3.5.   French-English mathematician Abraham de Moivre

the Central Limit Theorem, perhaps the most important theorem in probability theory.

Binet's formula can be proved by induction on n. When $n = 0$ or 1 there is nothing to prove. Suppose $n \geq 0$ and the results holds for all $1 \leq k \leq n$. The from $F_{n+1} = F_n + F_{n-1}$, we have

$$F_{n+1} = \frac{1}{\sqrt{5}} \left(\left(\frac{1 + \sqrt{5}}{2} \right)^n - \left(\frac{1 - \sqrt{5}}{2} \right)^n \right)$$

$$+ \frac{1}{\sqrt{5}} \left(\left(\frac{1 + \sqrt{5}}{2} \right)^{n-1} - \left(\frac{1 - \sqrt{5}}{2} \right)^{n-1} \right)$$

$$= \frac{1}{\sqrt{5}} \left(\left(\frac{3 + \sqrt{5}}{2} \right) \left(\frac{1 + \sqrt{5}}{2} \right)^{n-1} - \left(\frac{3 - \sqrt{5}}{2} \right) \left(\frac{1 - \sqrt{5}}{2} \right)^{n-1} \right)$$

$$= \frac{1}{\sqrt{5}} \left(\left(\frac{1 + \sqrt{5}}{2} \right)^{n+1} - \left(\frac{1 - \sqrt{5}}{2} \right)^{n+1} \right).$$

When $-n$ is a negative integer, we can use this result and the definition in the previous section of F_{-n} to obtain

$$F_{-n} = (-1)^{n-1} F_n$$

$$= \frac{(-1)^{n-1}}{\sqrt{5}} \left(\left(\frac{1 + \sqrt{5}}{2} \right)^n - \left(\frac{1 - \sqrt{5}}{2} \right)^n \right)$$

$$= \frac{1}{\sqrt{5}} \left(\left(\frac{1 + \sqrt{5}}{2} \right)^{-n} - \left(\frac{1 - \sqrt{5}}{2} \right)^{-n} \right),$$

so Binet's formula holds for all integers n. □

In 1728, the Swiss mathematician Nicolaus Bernoulli (1687–1759) used the method of generating functions to provide a different proof for Binet's formula. Here is Bernoulli's proof.

Consider the generating function

$$F(x) = \sum_{n=0}^{\infty} F_n x^n. \tag{3.3}$$

Then by an easy calculation,

$$F(x) - xF(x) - x^2 F(x) = x,$$

or

$$F(x) = \frac{x}{1 - x - x^2}.$$

(see Chapter 4, Section 7 for more details about this argument). Note that we can write

$$1 - x - x^2 = (1 - \alpha x)(1 - \beta x),$$

where $\frac{1}{\alpha}$ and $\frac{1}{\beta}$ are the two roots of the polynomial $1 - x - x^2$, that is $\frac{1 \pm \sqrt{5}}{2}$. Then we take partial fractions

$$\frac{x}{1 - x - x^2} = \frac{A}{1 - \alpha x} + \frac{B}{1 - \beta x},$$

where A and B satisfy $A + B = 0$, $A\beta + B\alpha = -1$; we solve these equations to find

$$A = \frac{1}{\alpha - \beta},$$

$$B = -\frac{1}{\alpha - \beta}.$$

Therefore

$$F(x) = \frac{1}{\alpha - \beta} \left(\frac{1}{1 - \alpha x} - \frac{1}{1 - \beta x} \right)$$

$$= \sum_{n=0}^{\infty} \frac{a^n - \beta^n}{\alpha - \beta} x^n. \tag{3.4}$$

Comparing the coefficients in (3.4) with (3.3), we have Binet's formula. □

From this formula, we can immediately derive that

$$\frac{F_{n+1}}{F_n} = \frac{\alpha}{1 - \left(\frac{\alpha}{\beta}\right)^n} - \frac{\beta}{\left(\frac{\beta}{\alpha}\right)^n - 1}.$$

Since

$$\left|\frac{\beta}{\alpha}\right| = \frac{3 - \sqrt{5}}{2} < 1,$$

$$\left|\frac{\alpha}{\beta}\right| = \frac{3 + \sqrt{5}}{2} > 1,$$

we can evaluate the limit of the sequence $\phi_n = \frac{F_{n+1}}{F_n}$.

Theorem 3.1. *The limit of the sequence ϕ_n is*

$$\lim_{n \to \infty} \phi_n = \frac{1 + \sqrt{5}}{2}.$$

This number is about $1.618\ldots$, *its reciprocal* $0.618\ldots$: *the so-called golden ratio.*

Incidentally, it was the the astronomer Johannes Kepler (1571–1630) who first observed the existence of this limit. The Binet formula was rediscovered in 1843 by Jacques Binet (1786–1856). After whom it is named. In 1844, the French mathematician Gabriel Lamé (1795–1870) also independently discovered this formula.

In his novel *The Da Vinci Code*, the American author Dan Brown (1964–) put beside the body of the Jacques Saunoère the numbers 13, −3, −2, −21, −1, −1, −8, −5 (Fig. 3.6). His granddaughter recognized these as Fibonacci numbers and used them to find out the code 1123481321 (consisting of the first eight Fibonacci numbers) to his safe. Also, in the preface to the English version of *A Modern Introduction to Classical Number Theory* (World Scientific Press, 2021) by the present author, the page numbers for the seven illustrations are 1, 1, 2, 3, 5, 8, 13.

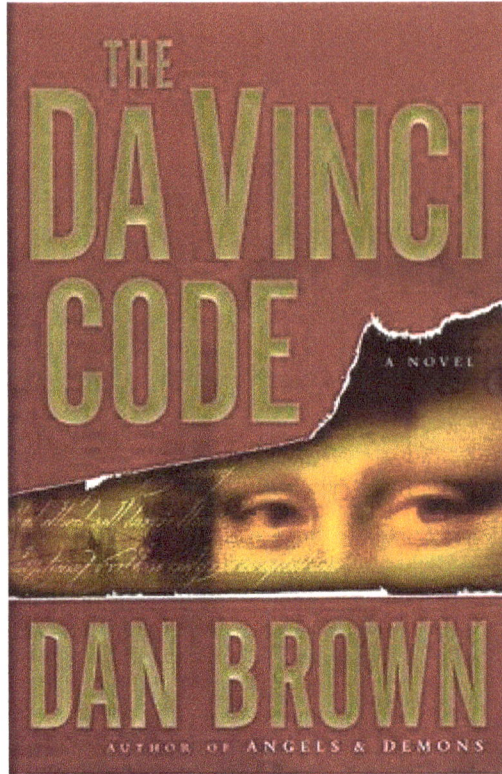

Figure 3.6.   *The Da Vinci Code*, 1st Edition (2003)

3.4.   Connection with Continued Fractions

Mathematicians have also known for a long time that the numbers ϕ_n can be represented by continued fractions. *Continued fractions* consist of finitely or infinitely many fractions with a predetermined pattern. Any rational number can be written as a continued fraction in two ways. For example,

$$\frac{10}{7} = 1 + \cfrac{1}{2 + \frac{1}{3}},$$

which we express compactly with the notation $\frac{10}{7} = [1; 2, 3]$. Also, $\frac{10}{7} = [1; 2, 2, 1]$.

The Fibonacci sequence produces a particularly special sequence of continued fractions,

$$\phi_n = 1 + \cfrac{1}{1 + \cfrac{1}{\ddots + \cfrac{1}{1}}},$$

$$\underbrace{\qquad\qquad\qquad}_{n \text{ ones}}$$

in which there occur a total of exactly n ones along the diagonal; or compactly $\phi_n = \underbrace{[1; 1, \ldots, 1]}_{n \text{ ones}}$ (Fig. 3.7). Corresponding to this, we have the infinite continued fraction

$$[1; 1, 1, \ldots] = 1.618\ldots.$$

If we express this as $[1; 1, 1, \ldots] = [1; \overline{1}]$, we can consider other periodic continued fractions; for example, $[1; \overline{2}] = \sqrt{2}$. In general, the irrational roots of any quadratic polynomial with integer coefficients admit an expression as a periodic continued fraction (see [5]). The above expression for $\sqrt{2}$ for example is derived as

$$\sqrt{2} = 1 + (\sqrt{2} - 1)$$

$$= 1 + \cfrac{1}{1 + \sqrt{2}}$$

$$= 1 + \cfrac{1}{1 + \left(1 + \cfrac{1}{1 + \sqrt{2}}\right)}$$

$$= 1 + \cfrac{1}{2 + \cfrac{1}{1 + \sqrt{2}}}$$

$$= 1 + \cfrac{1}{2 + \cfrac{1}{2 + \cfrac{1}{1 + \sqrt{2}}}}$$

$$\vdots$$

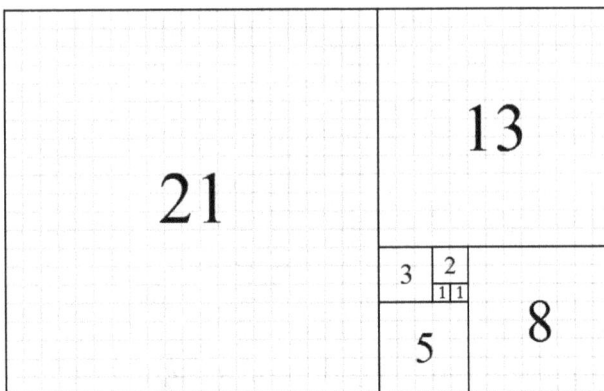

Figure 3.7.   A square fold with Fibonacci number sidelength

by which point the periodicity is already obvious. Similar arguments produce the periodic continued fractions

$$\sqrt{2} = [1; \overline{2}]$$
$$\sqrt{3} = [1; \overline{1, 2}]$$
$$\sqrt{5} = [2; \overline{4}]$$
$$\sqrt{7} = [2; 1, 1, 1, \overline{4}]$$

$$\vdots$$

We close this section with another approximation.

Theorem 3.2. *For all* $n \geq 11$,

$$\left(\frac{3}{2}\right)^n < F_n < \left(\frac{5}{3}\right)^n. \tag{3.5}$$

Moreover, the second inequality is valid for all $n \geq 0$.

Proof.   The proof is by induction on n. Consider first the second inequality; if $n = 0$ or 1, there is nothing to prove. Let $n \geq 2$ and suppose $F_k < (\frac{5}{3})^k$ for all $0 \le k \le n - 1$. Then

$$F_n = F_{n-1} + F_{n-2}$$

$$< \left(\frac{5}{3}\right)^{n-1} + \left(\frac{5}{3}\right)^{n-2}$$

$$= \left(\frac{5}{3}\right)^{n-2}\left(\frac{5}{3} + 1\right)$$

$$< \left(\frac{5}{3}\right)^n,$$

as required.

Next, we prove the first inequality. Note that

$$F_{11} = 89 > \left(\frac{3}{2}\right)^{11} \approx 86.5,$$

$$F_{12} = 144 > \left(\frac{3}{2}\right)^{12} \approx 129.7.$$

Let $n \ge 12$ and suppose the result holds for all $11 \le k \le n$. Then similarly

$$F_{n+1} = F_n + F_{n-1}$$

$$> \left(\frac{3}{2}\right)^n + \left(\frac{3}{2}\right)^{n-1}$$

$$= \left(\frac{3}{2}\right)^{n-1}\left(\frac{5}{2}\right)$$

$$> \left(\frac{3}{2}\right)^{n+1}.$$

This completes the proof. $\square$

We can also show by a similar argument that for all $n \ge 35$,

$$\left(\frac{10}{7}\right)^n < G_n < \left(\frac{11}{7}\right)^n.$$

3.5.   Three Identities

In 1680, the Italian (naturalized French) mathematician and astronomer Giovanni Cassini, after whom the Cassini Division in the rings of Saturn is named, discovered the following identity during his tenure as director of the Paris Observatory (Fig. 3.8):

$$F_{n-1}F_{n+1} - F_n^2 = (-1)^n,$$

which later came to be called Cassini's identity. Two centuries later, in 1876, Lucas contributed also the following identities, using induction and the Fibonacci recurrence:

$$\sum_{k=0}^{n} F_k = F_{n+2} - 1,$$

$$\sum_{k=0}^{n-1} F_{2k+1} = F_{2n},$$

Figure 3.8.   French astronomer Gian Domenico Cassini

$$\sum_{k=0}^{n} F_{2k} = F_{2n+1} - 1, \tag{3.6}$$

$$\sum_{k=0}^{n} F_k^2 = F_n F_{n+1},$$

$$F_{2n-1} = F_n^2 + F_{n-1}^2$$

$$F_{2n} = F_{n+1}^2 - F_{n-1}^2.$$

We shall make frequent use of Cassini's identity throughout the remainder of this book: one easy consequence of it is that adjacent Fibonacci numbers are relatively prime. First we present a proof, using a matrix argument given in 1997 by the American computer scientist and recipient of the 1997 Turing Award Donald Knuth (1938–) (it is also possible to prove it in the usual way by induction and the Fibonacci recurrence relation). The trick is to recognize the left-hand side as the determinant of a 2×2 matrix:

$$F_{n-1}F_{n+1} - F_n^2 = \begin{vmatrix} F_{n+1} & F_n \\ F_n & F_{n-1} \end{vmatrix}$$

$$= \left| \begin{pmatrix} 1 & 1 \\ 1 & 0 \end{pmatrix} \begin{pmatrix} F_n & F_{n-1} \\ F_{n-1} & F_{n-2} \end{pmatrix} \right|$$

$$\vdots$$

$$= \left| \begin{pmatrix} 1 & 1 \\ 1 & 0 \end{pmatrix}^{n-1} \begin{pmatrix} F_2 & F_1 \\ F_1 & F_0 \end{pmatrix} \right|$$

$$= \begin{vmatrix} 1 & 1 \\ 1 & 0 \end{vmatrix}^n$$

$$= (-1)^n.$$

In 1879, the French-Belgian mathematician Eugène Catalan (1814–1894) proved a generalization of the Cassini identity, now called the Catalan identity (Fig. 3.9):

$$F_n^2 - F_{n-r}F_{n+r} = (-1)^{n-r} F_r^2 \quad \text{for all } 1 \leq r \leq n.$$

So Cassini's identity is the special case $r = 1$.

Figure 3.9.   French mathematician Eugène Charles Catalan

Again, we recognize the left side as a determinant:

$$F_n^2 - F_{n-r}F_{n+r} = - \begin{vmatrix} F_{n+r} & F_n \\ F_n & F_{n-r} \end{vmatrix}.$$

By (3.2), we can rewrite this determinant as

$$\begin{vmatrix} F_{n+r} & F_n \\ F_n & F_{n-r} \end{vmatrix} = \begin{vmatrix} F_{r+1}F_n + F_rF_{n-1} & F_n \\ F_{r+1}F_{n-r} + F_rF_{n-r-1} & F_{n-r} \end{vmatrix}$$

$$= \begin{vmatrix} F_rF_{n-1} & F_n \\ F_rF_{n-r-1} & F_{n-r} \end{vmatrix}$$

$$= F_r \begin{vmatrix} F_{n-1} & F_n \\ F_{n-r-1} & F_{n-r} \end{vmatrix}. \tag{3.7}$$

If $n - r$ is even, we reduce this iteratively as follows:

$$\begin{vmatrix} F_{n-1} & F_n \\ F_{n-r-1} & F_{n-r} \end{vmatrix} = \begin{vmatrix} F_{n-1} & F_{n-2} \\ F_{n-r-1} & F_{n-r-2} \end{vmatrix}$$

$$\vdots$$

$$= \begin{vmatrix} F_{n-(n-r)} & F_{n-(n-r)+1} \\ F_{n-r-(n-r)} & F_{n-r-(n-r)+1} \end{vmatrix}$$

$$= \begin{vmatrix} F_{r+1} & F_r \\ F_1 & F_0 \end{vmatrix}$$

$$= \begin{vmatrix} F_{r+1} & F_r \\ 1 & 0 \end{vmatrix}$$

$$= -F_r.$$

If $n - r$ is odd, we have instead

$$\begin{vmatrix} F_{n-1} & F_n \\ F_{n-r-1} & F_{n-r} \end{vmatrix} = \begin{vmatrix} F_{n-1} & F_{n-2} \\ F_{n-r-1} & F_{n-r-2} \end{vmatrix}$$

$$\vdots$$

$$= \begin{vmatrix} F_{n-(n-r)} & F_{n-(n-r)} \\ F_{n-r-(n-r)+1} & F_{n-r-(n-r)} \end{vmatrix}$$

$$= \begin{vmatrix} F_r & F_{r+1} \\ F_0 & F_1 \end{vmatrix}$$

$$= \begin{vmatrix} F_r & F_{r+1} \\ 0 & 1 \end{vmatrix}$$

$$= F_r.$$

These results and (3.7) give Catalan's identity.

Finally, the 20th century mathematician Steven Vajda (1901–1995), who was born in Hungary and educated in Austria before fleeing the Nazi regime for Britain, generalized these results still further to

$$F_{n+i}F_{n+j} - F_nF_{n+i+j} = (-1)^n F_i F_j;$$

this is Vajda's identity. We prove it here using Binet's formula. Again let α, β respectively be $\frac{1\pm\sqrt{5}}{2}$. Then

$$F_{n+i}F_{n+j} - F_nF_{n+i+j} = \frac{\alpha^{n+i} - \beta^{n+i}}{\sqrt{5}} \cdot \frac{\alpha^{n+j} - \beta^{n+j}}{\sqrt{5}}$$

$$-\frac{\alpha^n - \beta^n}{\sqrt{5}} \cdot \frac{\alpha^{n+i+j} - \beta^{n+i+j}}{\sqrt{5}}$$

$$= \frac{1}{5}(\alpha^n \beta^{n+i+}j + \alpha^{n+i+j}\beta^n$$

$$- \alpha^{n+i}\beta^{n+j} - \alpha^{n+j}\beta^{n+i})$$

$$= \frac{1}{5}(\alpha\beta)^n \left(\beta^{i+j} + \alpha^{i+j} - \alpha^i\beta^j - \alpha^j\beta^i\right)$$

$$= (\alpha\beta)^n \frac{a^i - \beta^i}{\sqrt{5}} \cdot \frac{a^j - \beta^j}{\sqrt{5}}$$

$$= (-1)^n F_i F_j.$$

We present next two results with a similar flavor to the various identities we have just shown. Observe first that by the familiar properties of determinants

$$\begin{vmatrix} F_n & F_{n+1} & F_{n+2} \\ F_{n+1} & F_{n+2} & F_{n+3} \\ F_{n+2} & F_{n+3} & F_{n+4} \end{vmatrix} = 0,$$

for all integers n. The following theorem gives a generalization.

Theorem 3.3. *For all integers n, k, we have*

$$\begin{vmatrix} F_n & F_{n+k} & F_{n+2k} \\ F_{n+k} & F_{n+2k} & F_{n+3k} \\ F_{n+2k} & F_{n+3k} & F_{n+4k} \end{vmatrix} = 0.$$

Proof. The following identities all follow from (3.2), the Fibonacci recurrence, and the familiar properties of determinants:

$$\begin{vmatrix} F_n & F_{n+k} & F_{n+2k} \\ F_{n+k} & F_{n+2k} & F_{n+3k} \\ F_{n+2k} & F_{n+3k} & F_{n+4k} \end{vmatrix} = \begin{vmatrix} F_n & F_{n+k} & F_{n+2k} \\ F_{n+k} & F_{n+(k+k)} & F_{n+(k+2k)} \\ F_{n+2k} & F_{n+(2k+k)} & F_{n+(2k+2k)} \end{vmatrix}$$

$$= \begin{vmatrix} F_n & F_{k+1}F_n + F_k F_{n-1} & F_{2k+1}F_n + F_{2k}F_{n-1} \\ F_{n+k} & F_{k+1}F_{n+k} + F_k F_{n+k-1} & F_{2k+1}F_{n+k} + F_{2k}F_{n+k-1} \\ F_{n+2k} & F_{2k+1}F_{n+2k} + F_k F_{n+2k-1} & F_{2k+1}F_{n+2k} + F_{2k}F_{n+2k-1} \end{vmatrix}$$

$$= F_k F_{2k} \begin{vmatrix} F_n & F_{n-1} & F_{n-1} \\ F_{n+k} & F_{n+k-1} & F_{n+k-1} \\ F_{n+2k} & F_{n+2k-1} & F_{n+2k-1} \end{vmatrix} = 0. \qquad \square$$

3.6. Equations Between Binomial Coefficients

The binomial coefficients appear everywhere in mathematics; a common thread running through such diverse branches as analysis and combinatorics. In number theory, too, there are interesting questions to ask about them. Here we consider only identities between binomial coefficients, in which the Fibonacci numbers play a special role (Fig. 3.10). Recall first the trivial identities

$$\binom{n}{0} = \binom{n}{n} = 1,$$

$$\binom{n}{1} = n, \text{ or}$$

$$\binom{\binom{n}{k}}{1} = \binom{n}{k}, \text{ and}$$

$$\binom{n}{k} = \binom{n}{n-k},$$

for all $n \geq 1$ and $0 \leq k \leq n$. We would like to know about any nontrivial solutions to the equation

$$\binom{m}{k} = \binom{n}{l} \qquad (3.8)$$

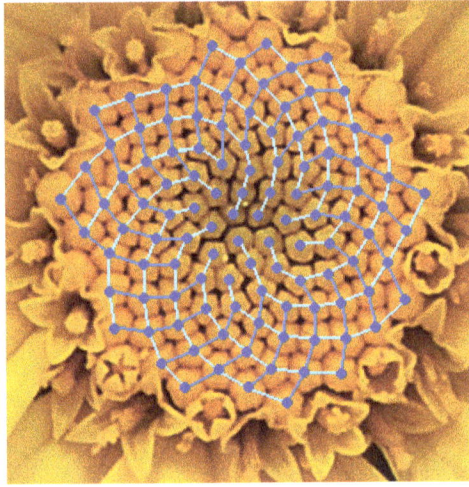

Figure 3.10. Yellow chamomile flower, the green and blue spirals have 13 and 21 nodes respectively

with $2 \le k \le m/2$, $2 \le l \le n/2$. Among numbers smaller than 10^{30} (or, when $\max(n, m) \le 1000$) the only solutions are

$$\binom{16}{2} = \binom{10}{3} = 120, \quad \binom{21}{2} = \binom{10}{4} = 210,$$

$$\binom{56}{2} = \binom{22}{3} = 1540, \quad \binom{120}{2} = \binom{36}{3} = 7140,$$

$$\binom{153}{2} = \binom{19}{5} = 11628, \quad \binom{221}{2} = \binom{17}{8} = 24310,$$

$$\binom{78}{2} = \binom{15}{5} = \binom{14}{6} = 3003,$$

and the solutions

$$\binom{F_{2t+2}F_{2t+3}}{F_{2t}F_{2t+3}} = \binom{F_{2t-2}F_{2t+3} - 1}{F_{2t}F_{2t+3} + 1},$$

with $t \ge 1$.

The final equation shows that the Fibonacci numbers generate infinitely many binomial coefficient identities. This equation was obtained independently by D.A. Lind (*Fibonacci Quarterly* 6. 1968,

86–93) and D. Singmaster (*Fibonacci Quarterly* 13. 1975, 295–298). When $t = 1$ and 2 respectively, we get

$$\binom{15}{5} = \binom{14}{6},$$

and

$$\binom{104}{39} = \binom{103}{40}.$$

To verify this identity, consider $A = F_{2t+2} F_{2t+3}$, $B = F_{2t} F_{2t+3}$. Then it is identical to

$$\frac{A!}{B!(A-B)!} = \frac{(A-1)!}{(B+1)!(A-B-2)!},$$

or

$$A(B+1) = (A-B)(A-B-1).$$

Replacing A and B again with their values in Fibonacci numbers, this is

$$F_{2t+2} F_{2t+3}(F_{2t} F_{2t+3} + 1) = F_{2t+1} F_{2t+3}(F_{2t+1} F_{2t+3} - 1)$$

or, canceling like terms and invoking the Fibonacci recurrence,

$$F_{2t} F_{2t+2} + 1 = F_{2t+1}^2,$$

which is just the Cassini identity. □

De Weger (*Journal of Number Theory*, 63. 1997, 373–386) has put forward the conjecture that these eight identities are the only solutions.

In *A Modern Introduction to Classical Number Theory*, we introduced the *figurate primes*, which are 1 and all numbers of the form $\binom{p^k}{j}$ with p a prime number and j, k positive integers satisfying $j \leq \frac{p^k}{2}$. It is obvious that the binomial coefficients of this form include all primes and prime powers. We put forward several conjectures concerning these numbers; among them Conjecture 3.1 is a variation on Goldbach's conjecture, and applies to both even and odd numbers.

Conjecture 3.1. *Every positive integer larger than 1 admits a presentation as a sum of two figurate primes.*

Conjecture 3.2. *Every binomial coefficient in the definition of the figurate primes gives a different number.*

None of the eight identities listed above include figurate primes, so Conjecture 3.2 follows from De Weger's conjecture. But we have not so far been able to prove either of these conjectures.

3.7.   Divisibility Sequences

A *divisibility sequence* is a sequence $(a_n)_{n\geq 1}$ such that a_m divides a_n whenever m divides n. If also $\gcd(a_m, a_n) = a_{\gcd(m,n)}$, then $(a_n)_{n\geq 1}$ is called a *strong divisibility sequence*. For example, every sequence of the form $(kn)_{n\geq 1}$ with k any nonzero integer and any sequence of the form $(A^n - B^n)_{n\geq 1}$ with $A > B > 0$ is a divisibility sequence; every constant sequence is a strong divisibility sequence.

We have seen already that the Fibonacci numbers F_n with $n \geq 1$ form a divisibility sequence. In 1876, Lucas proved that they form a strong divisibility sequence (Fig. 3.11).

Theorem 3.4 (Lucas). *For all positive integers m, n,*

$$\gcd(F_m, F_n) = F_{\gcd(m,n)}. \tag{3.9}$$

Proof.   We prove this via Euclidean division. First, consider the addition rule (3.2): for any integers k, n and r with n and r not both

Figure 3.11.   French mathematician Edouard Lucas, who first named the Fibonacci sequence

zero, we have

$$F_{kn+r} = F_{(k-1)n+r}F_{n+1} + F_{(k-1)n+r-1}F_n,$$

and since adjacent Fibonacci numbers are relatively prime, therefore

$$\gcd(F_{kn+r}, F_n) = \gcd(F_{(k-1)n+r}, F_n) = \gcd(F_{(k-1)n+1}F_{n+1}, F_n),$$

and continuing inductively,

$$\gcd(F_{kn+r}, F_n) = \gcd(F_{(k-1)n+r}, F_n)$$

$$\vdots$$

$$= \gcd(F_{n+r}, F_n)$$
$$= \gcd(F_r, F_n). \qquad (3.10)$$

Now assume without loss of generality that $m > n$, and use Euclidean division to find the numbers $r_0, \ldots, r_s$ with $0 < r_s < r_{s-1} < \cdots < r_1 < r_0 < n$ such that

$$m = nq + r_0,$$
$$n = r_0q_1 + r_1$$

$$\vdots$$

$$r_{s-2} = r_{s-1}q_s + r_s$$
$$r_{s-1} = r_sq_{s+1}.$$

In particular, $r_s = \gcd(m, n)$. Then by (3.10),

$$\gcd(F_m, F_n) = \gcd(F_{qn+r}, F_r) = \gcd(F_{qn+r}, F_n)$$
$$= \gcd(F_n, F_r) = \gcd(F_r, F_{r_1})$$

$$\vdots$$

$$= \gcd(F_{r_{s-1}}, F_{r_s}) = \gcd(F_{q_{s+1}r_s}, F_{r_s})$$
$$= \gcd(F_{r_s}, F_{r_s}) = F_{r_s}$$
$$= F_{\gcd(m,n)}.$$

$$\square$$

Corollary 3.1. F_m *divides* F_n *if and only if* m *divides* n.

Corollary 3.2. *If* $\gcd(m, n) = 1$, *then* $F_m F_n$ *divides* F_{mn}.

It follows that, with the exception of $F_4 = 3$, if F_n is prime, then n must be prime. Since there exist arbitrarily long strings of consecutive composite numbers, therefore also there exist arbitrarily long strings of consecutive composite Fibonacci numbers.

We can also use Lucas's theorem to give another proof that there are infinitely many prime numbers. Otherwise, suppose that $p_1, \ldots, p_k$ are all the primes. From Lucas's theorem, the Fibonacci numbers $F_{p_1}, \ldots, F_{p_k}$ are relatively prime in pairs, so each $F_{p_1}, \ldots, F_{p_k}$ has one and only one prime factor; in other words, each of $F_{p_1}, \ldots, F_{p_k}$ is prime; therefore it suffices to exhibit a single composite Fibonacci number F_p with p prime to prove that there are infinitely many primes: for example, $F_{19} = 4181 = 37 \times 113$.

In addition to Lucas's theorem, if m divides n, then we also have the following divisibility results:

- if $\frac{n}{m} \equiv 0 \pmod 4$, then F_m divides $F_{n\pm1} - F_1$;
- if $\frac{n}{m} \equiv 1 \pmod 4$, then F_m divides $F_{n\pm1} + F_{m-2}$;
- if $\frac{n}{m} \equiv 2 \pmod 4$, then F_m divides $F_{n\pm1} + (-1)^{m-1} F_1$;
- if $\frac{n}{m} \equiv 3 \pmod 4$, then F_m divides $F_{n\pm1} + (-1)^m F_{m-2}$;
- if $\frac{n}{m} \equiv 0 \pmod 4$, then F_m divides $F_{n\pm2} \mp F_2$;
- if $\frac{n}{m} \equiv 1 \pmod 4$, then F_m divides $F_{n\pm2} \pm F_{m-2}$;
- if $\frac{n}{m} \equiv 2 \pmod 4$, then F_m divides $F_{n\pm2} + (-1)^{m-1} F_2$;
- if $\frac{n}{m} \equiv 3 \pmod 4$, then F_m divides $F_{n\pm2} + (-1)^m F_{m-2}$;
- if $\frac{n}{m} \equiv 0 \pmod 4$, then F_m divides $F_{n\pm3} - F_3$;
- if $\frac{n}{m} \equiv 1 \pmod 4$, then F_m divides $F_{n\pm3} - F_{m-3}$;
- if $\frac{n}{m} \equiv 2 \pmod 4$, then F_m divides $F_{n\pm3} + (-1)^{m-1} F_3$;
- if $\frac{n}{m} \equiv 3 \pmod 4$, then F_m divides $F_{n\pm3} + (-1)^m F_{m-3}$;
- if $\frac{n}{m} \equiv 0 \pmod 4$, then F_m divides $F_{n\pm4} \mp F_4$;
- if $\frac{n}{m} \equiv 1 \pmod 4$, then F_m divides $F_{n\pm4} \pm F_{m-4}$;

- if $\frac{n}{m} \equiv 2 \pmod 4$, then F_m divides $F_{n\pm4} + (-1)^{m-1} F_4$;
- if $\frac{n}{m} \equiv 3 \pmod 4$, then F_m divides $F_{n\pm4} \pm (-1) F_{m-4}$.

We now introduce elliptic divisibility sequences. An *elliptic divisibility sequence* (*EDS*) is a sequence of integers satisfying a nonlinear recurrence relation generated by division polynomials on elliptic curves. Such sequences were first proposed in 1948 by the American mathematician Morgan Ward (1901–1963), a student of Eric Temple Bell (1883–1960), who wrote *The Men of Mathematics*. The topic of elliptic divisibility sequences was a relatively obscure one until the new millenium; it has since attracted more attention, due to its relation to elliptic curves, logic, and cryptography.

Concretely, an elliptic divisibility sequence $(W_n)_{n\geq1}$ is determined by four initial values W_1, W_2, W_3, W_4 satisfying $W_1 W_2 W_3 \neq 0$ and the recurrence relation

$$\begin{cases} W_{2n+1} W_1^3 = W_{n+2} W_n^3 - W_{n+1}^3 W_{n-1} & (n \geq 2), \\ W_{2n} W_2 W_1^2 = W_{n+2} W_n W_{n-1}^2 - W_n W_{n-2} W_{n+1}^2 & (n \geq 3). \end{cases}$$

It is not difficult to prove by induction that if W_1 divides W_2, W_3, and W_4, and also W_2 divides W_4, then $(W_n)_{n\geq1}$ is a sequence of integers; note in particular that every W_{2n} is a multiple of W_2. In fact, every elliptic divisibility sequence is a divisibility sequence: if m divides n, then W_m divides W_n.

Example 3.1. The sequence $1, 2, 3, \ldots$ of positive integers is an elliptic divisibility sequence.

Example 3.2. The sequence $1, 3, 8, 21, 55, 144, 377, 987$ of Fibonacci numbers with even indices is an elliptic divisibility sequence.

A basic property of divisibility sequences is that

$$W_{n+m} W_{n-m} W_r^2 = W_{n+r} W_{n-r} W_m^2 - W_{m+r} W_{m-r} W_n^2,$$

whenever $n > m > 3$.

In 1913, Carmichael also proved the following result.

Carmichael's Theorem. *With the exceptions of* 1, 8 *and* 144, *every Fibonacci number has a prime factor that does not divide any previous Fibonacci number.*

3.8.   Zeckendorf's Theorem

In 1972, Edouard Zeckendorf (1901–1983), a retired Belgian doctor and military officer, proved the following theorem, later named after him.

Theorem 3.5 (Zeckendorf). *Every positive integer admits a unique presentation as a sum of one or more distinct pairwise nonadjacent Fibonacci numbers, where at most one of $F_1 = F_2 = 1$ occurs in the sum.*

Such a presentation is called a *Zeckendorf representation*; for example, the Zeckendorf representations of 50 and 100 are $50 = 3+13+34$, $100 = 3 + 8 + 89$ (Fig. 3.12). If we relax the requirement that the Fibonacci numbers in the sum be nonadjacent, then the representation is no longer unique: for example, also $100 = 55 + 34 + 8 + 3 = 89+8+2+1$. We present here separately the existence and uniqueness proofs: the former can be proved by induction, the latter either by set theoretical considerations or by the additional rule given in the previous section. Some further new results and proofs in this section were developed in our number theory seminar.

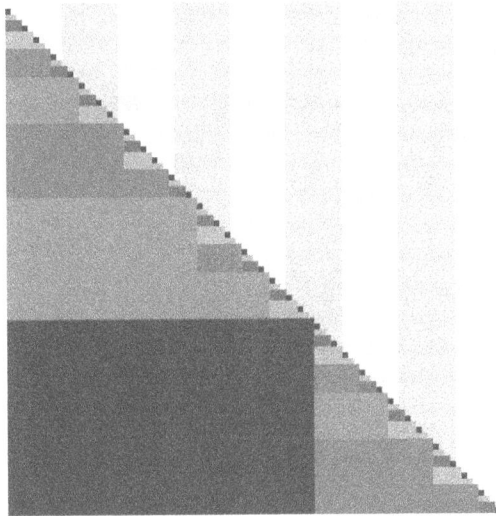

Figure 3.12.   Zeckendorf representations of numbers 1 through 89; the length and width of each rectangle are Fibonacci numbers

Proof of Theorem 3.5 (*Existence*). The numbers $n = 1, 2$ and 3 are Fibonacci numbers, and $4 = 1+3$. Suppose $n \geq 4$ the result holds for all $1 \leq k \leq n$. If $n + 1$ is a Fibonacci number, there is nothing to prove; otherwise, suppose $F_j < n + 1 < F_{j+1}$ and set $k = n + 1 - F_j$. Then $k \leq n$, so by the induction hypothesis we have a Zeckendorf representation $k = \sum$; also, since $k < F_{j+1} - F_j = F_{j-1}$, F_{j-1} does not appear as a term in $\sum$. It follows that $n + 1 = \sum + F_j$ is a Zeckendorf representation of $n + 1$, completing the induction.

(*Uniqueness*) Suppose for the sake of contradiction that some positive integer admits two Zeckendorf representations

$$F_{i_1} + \cdots + F_{i_s} = F_{j_1} + \cdots + F_{j_t}$$

with

$$i_1 > i_2 + 1 > \cdots > i_s + s - 1, \quad j_1 > j_2 + 1 > \cdots > j_t + t - 1. \quad (3.11)$$

We can assume without loss of generality that n is the smallest such integer, so $F_{i_1} \neq F_{j_1}$, say $F_{i_1} > F_{j_1}$ (otherwise $n - F_{i_1}$ would be a smaller such integer). Assume first that j_1 is even. Then from (3.6),

$$(3.11) \text{ RHS} \leq F_{j_1} + F_{j_1-2} + \cdots + F_2 = F_{j_1+1} - 1 < F_{j_1+1} \leq F_{i_1}.$$

On the other hand, if j_1 is odd, since one of F_1 or F_2 has been removed from the sum and the other is adjacent in the sum to F_3, then in the same way from (3.6) we get

$$(3.11) \text{ RHS} \leq F_{j_1} + F_{j_1-2} + \cdots + F_3 = F_{j_1+1} - 1 < F_{j_1+1} \leq F_{i_1}.$$

In either case, we have (3.11) RHS $<$ (3.11) LHS, which is a contradiction. □

Edouard Zeckendorf was born in Liège, a famous city in eastern Belgium. He graduated as a medical doctor from University of Liège and joined the Belgian army medical corps. In 1940, he was captured by the invading German army, and spent five years as a prisoner of war. Zeckendorf studied the Fibonacci sequence in his spare time, and published his theorem in 1972. In fact, the same result had been published twenty years earlier by the Dutch mathematician Gerrit Lekkerkerker.

We have the following variation of Zeckendorf's theorem.

Theorem 3.6. *Every positive integer admits a unique presentation as a sum of one or more of the numbers G_n, with each index in the sum separated from every other by at least 2; at most one of G_1, G_2, $G_3 = 1$ occurs in the sum.*

Existence is easy to prove along the lines of the previous proof; the uniqueness proof is also similar, but we need a version of (3.6) for the numbers G_n. By induction and the G_n recurrence relation, we can prove

$$\sum_{k=0}^{n} G_k = G_{n+3} - 1,$$

$$\sum_{k=0}^{n} G_{3k+1} = G_{3n+2},$$

$$\sum_{k=0}^{n} G_{3k+2} = G_{3n+3},$$

$$\sum_{k=0}^{n} G_{3k} = G_{3n+1} - 1.$$

We have not however been able to obtain properties similar to the sum of squares formula for the Fibonacci numbers or the greatest common divisor property for the sequence G_n. We do have versions of Cassini's and Catalan's identities, for example

$$G_{n-r}G_{n+r} - G_n^2 = g_r(n),$$

where

$$\begin{cases} g_r(r) = -G_r^2, \\ g_r(r-1) = -G_{r-1}^2, \\ g_r(n) = g_r(n-3) - g_r(n-2). \end{cases}$$

Also,

$$G_{2r} = G_{r+1}^2 + G_{r-1}^2 - G_{r-2}^2,$$

$$G_{2r+1} = G_{r+1}^2 + G_r^2 + G_{r-1}^2 - G_{r-3}^2 = G_{r+1}^2 + 2G_r G_{r-1}.$$

Finally, we have the following fifteen identities:

$$G_{n+1}G_{n+4} - G_{n+2}G_{n+3} = G_{-n-5},$$

$$G_{n+1}G_{n+5} - G_{n+2}G_{n+4} = G_{-n-5},$$

$$G_{n+1}G_{n+6} - G_{n+3}G_{n+4} = -G_{-n-7},$$

$$G_{n+1}G_{n+7} - G_{n+2}G_{n+6} = -G_{-n-10},$$

$$G_{n+1}G_{n+9} - G_{n+2}G_{n+8} = -2G_{-n-10},$$

$$G_{n+1}G_{n+10} - G_{n+2}G_{n+9} = -3G_{-n-10},$$

$$G_{n+1}G_{n+8} - G_{n+3}G_{n+6} = G_{-n-12},$$

$$G_{n+1}G_{n+10} - G_{n+3}G_{n+8} = 2G_{-n-12},$$

$$G_{n+1}G_{n+11} - G_{n+3}G_{n+9} = 3G_{-n-12},$$

$$G_{n+1}G_{n+9} - G_{n+4}G_{n+6} = G_{-n-12},$$

$$G_{n+1}G_{n+11} - G_{n+4}G_{n+8} = 2G_{-n-12},$$

$$G_{n+1}G_{n+12} - G_{n+4}G_{n+9} = 3G_{-n-12},$$

$$G_{n+1}G_{n+13} - G_{n+6}G_{n+8} = -2G_{-n-17},$$

$$G_{n+1}G_{n+14} - G_{n+6}G_{n+9} = -3G_{-n-17},$$

$$G_{n+1}G_{n+16} - G_{n+8}G_{n+9} = -6G_{-n-17}.$$

We consider next the determinant identity for any nonnegative integer n:

$$\begin{vmatrix} G_n & G_{n+1} & G_{n+2} \\ G_{n+1} & G_{n+2} & G_{n+3} \\ G_{n+2} & G_{n+3} & G_{n+4} \end{vmatrix} = -1. \qquad (3.12)$$

More generally, for any nonnegative n and $k > 0$, we consider the determinant

$$G(n,k) = \begin{vmatrix} G_n & G_{n+k} & G_{n+2k} \\ G_{n+k} & G_{n+2k} & G_{n+3k} \\ G_{n+2k} & G_{n+3k} & G_{n+4k} \end{vmatrix}. \qquad (3.13)$$

First, we prove (3.12). For $n = 0$, it can be easily verified by direct calculation. Suppose $n \geq 1$, and (3.12) holds for $n - 1$. Then by standard matrix operations and the G_n recurrence, we subtract the second row from the third to obtain

$$\begin{vmatrix} G_n & G_{n+1} & G_{n+2} \\ G_{n+1} & G_{n+2} & G_{n+3} \\ G_{n+2} & G_{n+3} & G_{n+4} \end{vmatrix} = \begin{vmatrix} G_n & G_{n+1} & G_{n+2} \\ G_{n+1} & G_{n+2} & G_{n+3} \\ G_{n-1} & G_n & G_{n+1} \end{vmatrix},$$

and by two row interchanges:

$$\begin{vmatrix} G_n & G_{n+1} & G_{n+2} \\ G_{n+1} & G_{n+2} & G_{n+3} \\ G_{n+2} & G_{n+3} & G_{n+4} \end{vmatrix} = \begin{vmatrix} G_{n-1} & G_n & G_{n+1} \\ G_n & G_{n+1} & G_{n+2} \\ G_{n+1} & G_{n+2} & G_{n+3} \end{vmatrix} = -1,$$

by the induction hypothesis. This proves (3.12).   □

We return to (3.13). By the previous result, we can replace $G(n, k)$ with simply $G(k)$. With $k = 1, \ldots, 10$, the first ten values of $G(k)$ are $-1, -1, -1, -9, -1, -121, -64, -729, -2809, -961$. Each of these is the negative of a square number, generally increasing to infinity in absolute value, but not monotonically.

Moving on, it is easy to see that the generating function for

$$G(x) = \sum_{n=0}^{\infty} G_n x^n,$$

is

$$G(x) = \frac{x}{1 - x^2 - x^3}.$$

Let α, β, γ be the three roots of $1 - x^2 - x^3$, say with α the real root, and with β, γ conjugate complex roots. By Rolle's theorem, $\alpha > 1$; we can also calculate α directly using the cubic formula:

$$\alpha = \sqrt[3]{\frac{1 + \sqrt{\frac{23}{27}}}{2}} + \sqrt[3]{\frac{1 - \sqrt{\frac{23}{27}}}{2}} \approx 1.32472.$$

Suppose the explicit formula for G_n has the expression

$$G_n = A\alpha^n + B\beta^n + C\gamma^n.$$

By Cramer's rule,

$$A = \frac{\alpha}{(\alpha - \beta)(\beta - \gamma)},$$

$$B = \frac{\beta}{(\beta - \gamma)(\gamma - \alpha)},$$

$$C = \frac{\gamma}{(\gamma - \alpha)(\alpha - \beta)}.$$

Then working with determinants,

$$G_n = \left| \begin{pmatrix} A & B & C \\ A\alpha^n & B\beta^n & C\gamma^n \\ A\alpha^{2n} & B\beta^{2n} & C\gamma^{2n} \end{pmatrix} \begin{pmatrix} 1 & \alpha^n & \alpha^{2n} \\ 1 & \beta^n & \beta^{2n} \\ 1 & \gamma^n & \gamma^{2n} \end{pmatrix} \right|$$

$$= ABC \left| \begin{matrix} 1 & 1 & 1 \\ \alpha^n & \beta^n & \gamma^n \\ \alpha^{2n} & \beta^{2n} & \gamma^{2n} \end{matrix} \right|^2$$

$$= ABC(\alpha^n - \beta^n)^2(\beta^n - \gamma^n)^2(\gamma^n - \alpha^n)^2.$$

For example,

$$G_1 = ABC(\alpha - \beta)^2(\beta - \gamma)^2(\gamma - \alpha)^2 = 1.$$

Combining these results,

$$G_n = -\left(\frac{(\alpha^n - \beta^n)(\beta^n - \gamma^n)(\gamma^n - \alpha^n)}{(\alpha - \beta)(\beta - \gamma)(\gamma - \alpha)} \right)^2.$$

If we compare this with Binet's formula

$$F_n = \frac{1}{\sqrt{5}} \left(\left(\frac{1 + \sqrt{5}}{2} \right)^n - \left(\frac{1 - \sqrt{5}}{2} \right)^n \right) = \frac{\alpha_1^n - \beta_1^n}{\alpha_1 - \beta_1},$$

where α_1, β_1 are $\frac{1 \pm \sqrt{5}}{2}$, it is natural to speculate about more general formulations.

Finally, we calculate

$$\lim_{n \to \infty} \frac{G_{n+1}}{G_n} = \alpha \approx 1.32472.$$

3.9. From Base 2 to Base 3

So far we have considered the recurrences $F_n = F_{n-1} + F_{n-2}$ and $G_n = G_{n-1} + G_{n-3}$. If we work in the opposite direction, we can consider also $E_n = E_{n-1} + E_{n-1}$ $(n > 1)$. If we set $E_1 = 1$, then this is just $E_n = 2^{n-1}$. From the familiar system of binary representation, it is immediately obvious that every positive integer can be represented as a sum of one or more distinct E_n, with no additional conditions on the separations of the indices. Then we consider the Fibonacci numbers F_n as a variation of the binary number system determined by the E_n; in this sense, Zeckendorf's theorem comes as no surprise!

We would like to establish a similar result by analogy with the base 3 number system; that is, can we find a nontrivial recursive sequence such that every positive integer admits a unique presentation as a sum of elements in this sequence, with no element occurring more than twice in the sum? If we consider the relationship between E_n and F_n, an obvious candidate sequence is:

$$H_0 = -1,$$
$$H_1 = 1,$$
$$H_n = H_{n-2} + 2H_{n-1} \ (n \geq 2).$$

The first ten terms of this sequence are 1, 1, 3, 7, 17, 41, 99, 239, 577, 1393, and it is easy to prove the following identities by induction:

$$2\sum_{k=1}^{n} H_k = H_{n+2} - H_{n+1},$$
$$2\sum_{k=1}^{n} H_{2k+1} = H_{2n} + 1,$$
$$2\sum_{k=1}^{n} H_{2k} = H_{2n+1} - 1,$$
$$2\sum_{k=1}^{n} H_k^2 = H_n H_{n+1} + 1.$$

(3.14)

Theorem 3.7. *Every positive integer m admits a unique presentation as a sum of the numbers H_n, in which each H_n occurs at most* *twice. If some H_n occurs twice in the sum, then H_{n-1} is not among the terms of the sum.*

Note that the conditions in Theorem 3.7 are consistent with the conditions of Zeckendorf's theorem for Fibonacci numbers.

Proof of Theorem 3.7 (*Existence*). It is obviously true when $m = 1, 2, 3$. Suppose $m \geq 3$ and the result holds for all $1 \leq k \leq m$. If $m+1$ is among the H_n, there is nothing to prove. Otherwise, consider $H_j < m + 1 < H_{j+1}$ and set $a = m + 1 - H_j$. Evidently $a \leq m$ and therefore admits a presentation $a = \sum$ of the required form. Also $H_j + a < H_{j+1} = 2H_j + H_{j-1}$, so $a < H_j + H_{j-1}$. It follows that H_j and H_{j-1} do not both appear in the presentation $a = \sum$, and therefore $m + 1 = \sum + H_j$ is a sum of the required for for $m + 1$.

(*Uniqueness*) Suppose for the sake of contradiction that some positive integer admits two distinct presentations of the required form, say

$$\alpha_1 H_{i_1} + \cdots + \alpha_s H_{i_s} = \beta_1 H_{j_1} + \cdots + \beta_t H_{j_t}, \qquad (3.15)$$

with every $\alpha_l, \beta_m = 1$ or 2; also $i_1 > i_2$, and if $\alpha_1 = 2$, then $i_1 > i_2+1$. Suppose without loss of generality that $i_1 > j_1$. If j_1 is even, then by (3.14)

$$(3.15) \text{ RHS} < 2H_{j_1} + 2H_{j_1-2} + \cdots + 2H_2 = H_{j_1+1} - 1.$$

Similarly, if j_1 is odd, then

$$(3.15) \text{ RHS} < 2H_{j_1} + 2H_{j_1-2} + \cdots + 2H_3 = H_{j_1+1} - 1.$$

In both cases then

$$(3.15) \text{ RHS} < H_{j_1+1} \leq H_{i_1} \leq (3.15) \text{ LHS},$$

which is a contradiction. $\square$

We also have an identity similar to the Vajda identity for the sequence of H_n:

$$H_{n+i}H_{n+j} - H_n H_{n+i+j} = (-1)^n \frac{(H_i + H_{i+1})(H_j + H_{j+1})}{2}. \qquad (3.16)$$

We prove this by the familiar method of generating functions. Set

$$H(x) = \sum_{n=0}^{\infty} H_n x^n$$

with generating function $H(x) = x^2 - 2x - 1$, and let α and β be the roots of this polynomial; evidently $\alpha\beta = -1$. We assume the explicit formula for H_n has the form $H_n = A\alpha^n + B\beta^n$. Then

$$H_{n+i}H_{n+j} - H_n H_{n+i+j} = - \begin{vmatrix} H_n & H_{n+1} \\ H_{n+j} & H_{n+i+j} \end{vmatrix}$$

$$= - \left| \begin{pmatrix} A\alpha^n & B\beta^n \\ A\alpha^{n+j} & B\beta^{n+j} \end{pmatrix} \begin{pmatrix} 1 & \alpha^i \\ 1 & \beta^i \end{pmatrix} \right|$$

$$= -(-1)^n AB(\alpha^i - \beta^i)(\alpha^j - \beta^j).$$

If we let the above expression be $H(n; i, j)$, then it is independent of n, and

$$H(0; 1, 1) = AB(\alpha - \beta)^2 = -2.$$

So we can solve for AB and substitute to get

$$H_{n+i}H_{n+j} - H_n H_{n+i+j} = 2\,(-1)^n \frac{(\alpha^i - \beta^i)(\alpha^j - \beta^j)}{(\alpha - \beta)^2}. \qquad (3.17)$$

On the other hand, since for any positive integer n, H_n and H_{n+1} are linearly independent solutions of a second-order linear difference equation, so any third solution $\frac{\alpha^n - \beta^n}{\alpha - \beta}$ must be a linear combination of them, that is

$$\frac{\alpha^n - \beta^n}{\alpha - \beta} = aH_n + bH_{n+1}.$$

By checking $n = 0$ and 1, we get that $a = b = \frac{1}{2}$; substituting this value in (3.17), we obtain (3.16). $\qquad \square$

The Fibonacci Product. Let a and b be positive integers with Zeckendorf representations

$$a = \sum_{i=0}^{k} F_{m_i},$$

$$b = \sum_{j=0}^{l} F_{n_j},$$

respectively. Then we define the *Fibonacci product* of a and b to be

$$a \circ b = \sum_{i=0}^{k} \sum_{j=0}^{l} F_{m_i + n_j}.$$

Table 3.1. Fibonacci coding.

N	Zeckendorf representation	Fibonacci code
1	F_2	11
2	F_3	011
3	F_4	0011
4	$F_2 + F_4$	1011
5	F_5	00011
6	$F_2 + F_5$	10011
7	$F_3 + F_5$	01011
8	F_6	000011
9	$F_2 + F_6$	100011
10	$F_3 + F_6$	010011
11	$F_4 + F_6$	001011
12	$F_2 + F_4 + F_6$	101011

For example, from $2 = F_2$, $4 = F_4 + F_2$ (we leave out F_1), we get

$$2 \circ 4 = F_{3+4} + F_{3+2} = 13 + 5 = 18.$$

It can be shown that the Fibonacci product is both associative and commutative.

In computer science, there is also the Fibonacci code, which is closely connected with Zeckendorf's theorem. *Fibonacci coding* is defined as follows: for a positive integer N, suppose F_n is the largest Fibonacci in the Zeckendorf representation of N; then the Fibonacci code for N consists of n bits, with both the nth and $(n-1)$th bits equal to 1. For every other F_k in the Zeckendorf representation of N, the $(k-1)$th bit is also 1, and every other bit is equal to 0. It follows from Zeckendorf's theorem that every positive integer has a unique Fibonacci code. Table 3.1 lists the Fibonacci codes for the first 12 positive integers. It is easy to see that the variations on Zeckendorf's theorem that we have presented above generate similar encodings.

3.10. Hilbert's Tenth Problem

In 1900, the German mathematician David Hilbert (1862–1943) put forward 23 famous mathematical problems for the new century at the International Congress of Mathematicians in Paris (Fig. 3.13).

Figure 3.13.   German mathematician David Hilbert

Figure 3.14.   Russian mathematician Yuri Matiyasevich

The tenth of these problems (Hilbert's tenth problem) was about the solvability in integers of Diophantine equations; specifically,

> Given a Diophantine equation with any number of unknown quantities and with rational integral numerical coefficients: To devise a process according to which it can be determined in a finite number of operations whether the equation is solvable in rational integers.

In 1970, the Russian mathematician Yuri Matiyasevich (1947–) proved that this problem has no solution. First we have the following theorem (Fig. 3.14).

Theorem 3.8. *If F_k^2 divides F_n, then F_k divides n.*

For example, $F_4^2 = 9$ divides $F_{12} = 144$ and $F_4 = 3$ divides 12; on the other hand, the converse is false in general: 5 divides 15, but $F_5^2 = 25$ does not divide $F_{15} = 610$.

Proof of Theorem 3.8. First, we prove by induction that

$$F_{kn+r} = \sum_{j=0}^{n} \binom{n}{j} F_{k-1}^{n-j} F_k^j F_{r+j},\qquad(3.18)$$

for every positive integer n. When $n = 1$, this is $F_{k+r} = F_{k-1}F_r + F_k F_{r+1}$, i.e. (3.2). Suppose the result is valid for some $n \geq 1$. Then

$$F_{k(n+1)+r} = F_{kn+(k+r)}$$

$$= \sum_{j=0}^{n} \binom{n}{j} F_{k-1}^{n-j} F_k^j F_{k+r+j}$$

$$= \sum_{j=0}^{n} \binom{n}{j} F_{k-1}^{n-j} F_k^j \left(F_k F_{r+j+1} + F_{k-1}F_{r+j} \right)$$

$$= \sum_{j=0}^{n} \binom{n}{j} F_{k-1}^{n-j} F_k^{j+1} F_{r+j+1}$$

$$+ \sum_{j=0}^{n} \binom{n}{j} F_{k-1}^{n-j+1} F_k^j F_{r+j}$$

$$= \sum_{j=1}^{n+1} \binom{n}{j} F_{k-1}^{n-j+1} F_k^j F_{r+j}$$

$$+ \sum_{j=0}^{n} \binom{n}{j} F_{k-1}^{n-j+1} F_k^j F_{r+j}$$

$$= \sum_{j=0}^{n+1} \binom{n+1}{j} F_{k-1}^{n-j+1} F_k^j F_{r+j},$$

as required; here we have used the identity (3.2) and the binomial coefficient identity

$$\binom{n+1}{j} = \binom{n}{j-1} + \binom{n}{j}.$$

Now suppose that some F_k^2 divides some F_n. Putting $r = 0$ in (3.18), we have

$$F_{kn} \equiv nF_{k-1}^{n-1}F_k \pmod{F_k^2},$$

from which it follows from the hypothesis that

$$nF_{k-1}^{n-1} \equiv 0 \pmod{F_k}.$$

Since adjacent Fibonacci numbers are relatively prime, this implies

$$n \equiv 0 \pmod{F_k},$$

as required.   □

From this, Matiyasevish was able to prove at the age of 23, while he was still a doctoral student at Leningrad State University (now Saint Petersburg State University), that the set $\{F_{2n}\}$ of Fibonacci numbers with even index forms a Diophantine set, and confirm an earlier hypothesis due to Julia Robinson called the *J.R. hypothesis*. This result established a negative answer to Hilbert's tenth problem. It is worth mentioning that the proof also made use of the Chinese Remainder Theorem "in a most extraordinary way".

At that time, research into Hilbert's tenth problem had already seen the following advances due to the work of Robinson, Martin Davis, and Hilary Putnam:

(1) every Diophantine set is recursively enumerable,
(2) every recursively enumerable set is exponential Diophantine
(3) conjecture: every exponential Diophantine set is a Diophantine set.

Matiyasevich completed the proof that every recursively enumerable set is Diophantine; that is, the recursively enumerable sets and the Diophantine sets are one and the same. This is known today as the Matiyasevich theorem, or the MRDP theorem.

A *Diophantine set* is a collection of all nonnegative integers (or lists of such integers) that make a Diophantine equation with parameters solvable in nonnegative integers when the values of its parameters are drawn from that collection. For example, it follows from the MRDP theorem that the set of prime numbers is Diophantine. More concretely, consider the Pell equation

$$x^2 - d(y+1)^2 = 1.$$

This is a Diophantine equation with parameter d with nonnegative integer solutions for the unknowns x and y whenever $d = 0$ or d is not a perfect square. This generates the Diophantine set

$$\{0, 2, 3, 5, 6, 7, 8, 10, 11, 12, 13, 14, 15, 17, \dots\}.$$

Some more examples: the equation $a = (2x + 3)y$ has nonnegative integer solutions if and only if a is not a power of 2; the equation $a = (x + 2)(y + 2)$ has nonnegative integer solutions if and only if $a \leq 3$ is composite; the equation $a = x + b$ generates the set of pairs (a, b) with $a \leq b$.

Hilbert's tenth problem serves as a bridge connecting number theory to computer science; in general, its negative solution is a powerful tool with which to establish the unsolvability of other propositions. Like Zeckendorf's theorem, it also has applications in cryptography.

In 1986, Matiyasevich and the British mathematician Richard Guy (1916–2020) discovered and proved (*American Mathematical Monthly*, 93.3, 631–535) that π can be expressed as the limit of a sequence involving Fibonacci numbers:

$$\pi = \lim_{n \to \infty} \sqrt{\frac{6 \log F_1, \dots, F_n}{\log \operatorname{lcm}(F_1, \dots, F_n)}}.$$

Chapter 4

Lucas Numbers and Lucas Sequences

> ... a servant, who found himself
> behind the sear of M. Édouard
> Lucas, dropped, by
> unskillfulness, a pile of plates.
> A broken piece of porcelain
> came to hit the cheek of M.
> Lucas and caused him a deep
> injury ... He took to his bed
> and soon appeared crysipelas
> which would take him away.
>
> An account of the death of
> Édouard Lucas in *La Lanterne*,
> October 1891

4.1. The Lucas Numbers

The *Lucas numbers* were first defined by the French mathematician
Édouard Lucas, whom we have already encountered more than once.
They are

$$\begin{cases} L_0 = 2, \\ L_1 = 1, \\ L_n = L_{n-1} + L_{n-1} \quad (n \geq 2). \end{cases}$$

The first 12 terms of this sequence (starting from the zeroth term) are

$$2, 1, 3, 4, 7, 11, 18, 29, 47, 76, 123, 199.$$

Like the Fibonacci sequence, we can extend the Lucas numbers to negative indices by

$$L_{-n} = (-1)^n.$$

The Lucas numbers and the Fibonacci numbers have many similar properties and relationships between them (Figs. 4.1 and 4.2); for example,

$$L_n = F_{n-1} + F_{n+1} = F_{n+2} - F_{n-2},$$
$$5F_n = L_{n+1} + L_{n-1} = \frac{1}{2}(L_{n+3} + L_{n-3}).$$

These formulas are easily proved by induction, We also have an explicit expression for the Lucas numbers:

$$L_n = \left(\frac{1 + \sqrt{5}}{2}\right)^n + \left(\frac{1 - \sqrt{5}}{2}\right)^n.$$

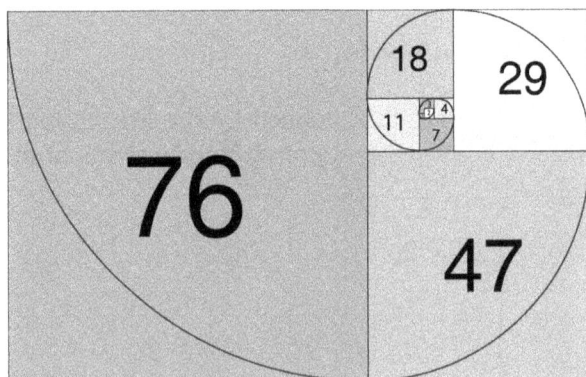

Figure 4.1.   The Lucas spiral

This follows immediately from Binet's formula and the above identities,

$$L_n = F_{n+1} + F_{n-1} = 2F_{n+1} - F_n$$

$$= \frac{1}{\sqrt{5}}\left(2\left(\frac{1+\sqrt{5}}{2}\right)^{n+1} - 2\left(\frac{1-\sqrt{5}}{2}\right)^{n+1}\right.$$

$$\left. - \left(\frac{1+\sqrt{5}}{2}\right)^{n} + \left(\frac{1-\sqrt{5}}{2}\right)^{n}\right)$$

$$= \left(\frac{1+\sqrt{5}}{2}\right)^{n} + \left(\frac{1-\sqrt{5}}{2}\right)^{n}.$$

On the other hand, if we had derived first the explicit expression for the Lucas numbers, then we can work backwards to obtain the Binet formula:

$$F_n = \frac{1}{5}(L_{n+1} + L_{n-1}) = \frac{1}{5}(L_n + 2L_{n-1})$$

$$= \frac{1}{5}\left(\left(\frac{1+\sqrt{5}}{2}\right)^{n} + 2\left(\frac{1+\sqrt{5}}{2}\right)^{n-1} + \left(\frac{1-\sqrt{5}}{2}\right)^{n}\right.$$

$$\left. + 2\left(\frac{1-\sqrt{5}}{2}\right)^{n-1}\right)$$

$$= \frac{1}{\sqrt{5}}\left(\left(\frac{1+\sqrt{5}}{2}\right)^{n} - \left(\frac{1-\sqrt{5}}{2}\right)^{n}\right).$$

Here we list various additional identities between Lucas numbers and Fibonacci numbers:

$$L_n^2 - L_{n-r}L_{n+r} = (-1)^n F_r^2,$$
$$L_{n+i}L_{n+j} - L_n L_{n+i+j} = (-1)^{n-1} \cdot 5F_i F_j,$$
$$F_{2n} = F_n L_n \quad \text{and} \quad 2F_{m+n} = F_m L_n + F_n L_m, \tag{4.1}$$
$$F_{n+k} + (-1)^k F_{n-k} = L_k F_n,$$

$$L_{n+k} + (-1)^k L_{n-k} = L_n L_k,$$

$$L_{m+n} = L_{m+1}F_n + L_m F_{n-1},$$

$$L_n^2 - 5F_n^2 = (-1)^n \, \dot{4}, \qquad (4.2)$$

$$\sum_{k=0}^{n} F_{2k} = F_{2n+1} - F_{-1} \quad \text{and} \quad \sum_{k=0}^{n} L_{2k} = L_{2n+1} - L_{-1};$$

$$\sum_{k=0}^{n} F_{2k-1} = F_{2n} - F_0 \quad \text{and} \quad \sum_{k=0}^{n} L_{2k-1} = L_{2n} - L_0, \qquad (4.3)$$

$$\sum_{k=0}^{n} (-1)^k F_{2k+1} = (-1)^n F_{n+1}^2 \quad \text{and}$$

$$\sum_{k=0}^{n} (-1)^k L_{2k+1} = (-1)^n F_{2n+2},$$

$$\sum_{k=0}^{\left\lfloor \frac{n+1}{2} \right\rfloor} (-1)^{\left\lfloor \frac{k+1}{2} \right\rfloor} F_{3k+2} = (-1)^{\left\lfloor \frac{n+1}{2} \right\rfloor} F_{\left\lfloor \frac{3n+1}{2} \right\rfloor}^2 \quad \text{and}$$

$$\sum_{k=0}^{\left\lfloor \frac{n+1}{2} \right\rfloor} (-1)^{\left\lfloor \frac{k+1}{2} \right\rfloor} L_{3k+2} = (-1)^{\left\lfloor \frac{n+1}{2} \right\rfloor} F_{2 \left\lfloor \frac{3n+4}{2} \right\rfloor},$$

$$\sum_{k=0}^{\left\lfloor \frac{n}{2} \right\rfloor} (-1)^{\left\lfloor \frac{k+1}{2} \right\rfloor} F_{3k+1} = (-1)^{\left\lfloor \frac{n}{2} \right\rfloor} F_{\left\lfloor \frac{3n}{2} \right\rfloor}^2 \quad \text{and}$$

$$\sum_{k=0}^{\left\lfloor \frac{n}{2} \right\rfloor} (-1)^{\left\lfloor \frac{k+1}{2} \right\rfloor} L_{3k+1} = (-1)^{\left\lfloor \frac{n}{2} \right\rfloor} F_{2 \left\lfloor \frac{3n+3}{2} \right\rfloor},$$

$$\sum_{k=1}^{n} L_k^2 = L_{2n+1} - L_0 + (-1)^n,$$

$$\sum_{k=0}^{n} \binom{n}{k} F_{k+r} = F_{2n+r} \quad \text{and} \quad \sum_{k=0}^{n} \binom{n}{k} L_{k+r} = L_{2n+r},$$

$$\sum_{k=0}^{n} \binom{n}{k} 2^k F_{k+r} = F_{3n+r} \quad \text{and} \quad \sum_{k=0}^{n} \binom{n}{k} 2^k L_{k+r} = L_{2n+r},$$

$$\sum_{k=0}^{n} (-1)^k \binom{n}{k} F_{2k+r} = (-1)^n F_{n+r} \quad \text{and}$$

$$\sum_{k=0}^{n} (-1)^k \binom{n}{k} L_{2k+r} = (-1)^n L_{n+r}.$$

We have the limits

$$\frac{L_n}{F_n} \xrightarrow[n\to\infty]{} \sqrt{5} \quad \text{and} \quad \frac{L_{n+1}}{L_n} \xrightarrow[n\to\infty]{} \frac{\sqrt{5}-1}{2}.$$

Finally, for any $0 \le t \le 3$,

$$\sum_{k=1}^{n} F_{4k-t} = F_{2n} F_{2n-t+2},$$

$$\sum_{k=1}^{n} L_{4k-t} = F_{4n-t+2} - F_{-t+2}.$$

Of course it is possible to find many more and very different forms, for example

$$F_n^2 - F_{n-2}^2 = F_{2n-2} \quad \text{and} \quad L_n^2 - L_{2n} = (-1)^n \cdot 2.$$

The following theorem is the analogue of Zeckendorf's theorem for the Lucas numbers.

Theorem 4.1. *Every positive integer admits a presentation as a sum of pairwise nonadjacent Lucas numbers, in which the numbers $L_0 = 2$ and $L_2 = 3$ do not both appear.*

The proof follows exactly the proof of Zeckendorf's theorem, using equation (4.3).

Next, we consider the divisibility properties of the Lucas numbers. Unlike the Fibonacci numbers, the Lucas numbers do not form a divisibility sequence. Nevertheless, we have the following results. Suppose m divides n. Then

- if $\frac{n}{m} \equiv 1 \pmod 2$, then L_m divides L_n,
- if $\frac{n}{m} \equiv 0 \pmod 4$, then L_m divides $L_n - 2$,
- if $\frac{n}{m} \equiv 2 \pmod 4$, then L_m divides $L_n + 2(-1)^m$.

The first of these was first proved in 1964 by the American mathematician Leonard Carlitz (1907–1999) (*Fibonacci Quarterly*, 1:2, Feb., 15–28). It follows that if m divides n, then L_m divides L_n if and only if $\frac{m}{n}$ is odd. For example, $L_4 = 7$ divides $L_{12} = 322$, but L_4 does not divide $L_8 = 47$. We have the following additional divisibility results:

- if $\frac{n}{m} \equiv 0 \pmod 4$, then L_m divides $L_{n\pm1} \mp L_1$,
- if $\frac{n}{m} \equiv 1 \pmod 4$, then L_m divides $L_{n\pm1} + L_{m-2}$,
- if $\frac{n}{m} \equiv 2 \pmod 4$, then L_m divides $L_{n\pm1} \pm (-1)^m L_1$,
- if $\frac{n}{m} \equiv 3 \pmod 4$, then L_m divides $L_{n\pm1} + (-1)^{m-1} L_{m-2}$,
- if $\frac{n}{m} \equiv 0 \pmod 4$, then L_m divides $L_{n\pm2} - L_2$,
- if $\frac{n}{m} \equiv 1 \pmod 4$, then L_m divides $L_{n\pm2} \pm L_{m-2}$,
- if $\frac{n}{m} \equiv 2 \pmod 4$, then L_m divides $L_{n\pm2} + (-1)^m L_2$,
- if $\frac{n}{m} \equiv 3 \pmod 4$, then L_m divides $L_{n\pm2} + (-1)^{m-1} L_{m-2}$,
- if $\frac{n}{m} \equiv 0 \pmod 4$, then L_m divides $L_{n\pm3} \mp L_3$,
- if $\frac{n}{m} \equiv 1 \pmod 4$, then L_m divides $L_{n\pm3} - L_{m-3}$,
- if $\frac{n}{m} \equiv 2 \pmod 4$, then L_m divides $L_{n\pm3} \pm (-1)^m L_3$,
- if $\frac{n}{m} \equiv 3 \pmod 4$, then L_m divides $L_{n\pm3} + (-1)^{m-1} L_{m-3}$,
- if $\frac{n}{m} \equiv 0 \pmod 4$, then L_m divides $L_{n\pm4} - L_4$,
- if $\frac{n}{m} \equiv 1 \pmod 4$, then L_m divides $L_{n\pm4} \pm L_{m-4}$,
- if $\frac{n}{m} \equiv 2 \pmod 4$, then L_m divides $L_{n\pm4} + (-1)^m L_4$,
- if $\frac{n}{m} \equiv 3 \pmod 4$, then L_m divides $L_{n\pm4} + (-1)^{m-1} L_{m-4}$.

Carlitz is also associated with the Carlitz–Wan conjecture, concerning the classification of certain polynomials over finite fields; this conjecture was introduced in its general form by Daqing Wan in 1993, and proved in 1995 by Hendrik Lenstra. In 1964, Carlitz also proved the following theorem.

Theorem 4.2. *The Lucas number L_m divides the Fibonacci number F_n if and only if $2m$ divides n.*

We have the following improvement upon this result.

Theorem 4.3. *Let n be a positive integer. If k is odd, then*

$$F_{2kn-2} \equiv (-1)^n \qquad (\mathrm{mod}\ L_n),$$
$$F_{2kn-1} \equiv (-1)^{n-1} \qquad (\mathrm{mod}\ L_n),$$

If k is even, then

$$F_{2kn-2} \equiv -1 \quad (\mathrm{mod}\ L_n),$$
$$F_{2kn-1} \equiv 1 \quad (\mathrm{mod}\ L_n),$$

By the Fibonacci recurrence, it is not hard to see that Theorem 4.2 is a corollary of Theorem 4.3. Theorem 4.3 can be derived from Theorem 3.2.13 in [11].

Next we consider an analogue to the relationship between the sequences G_n and F_n for the Lucas numbers. Define the sequence M_n by

$$\begin{cases} M_0 = 3, \\ M_1 = M_2 = 1, \\ M_n = M_{n-1} + M_{n-3}\ (n \geq 3); \end{cases}$$

the first several terms are 3, 1, 1, 4, 5, 6, 10, 15, 21, 31, 46, 67, 98, 144, 211, 309, 453, ... It is easy to prove by induction that

$$\sum_{k=0}^{n} M_k = M_{n+3} - 1,$$

$$\sum_{k=0}^{n} M_{3k+1} = M_{3n+2},$$

$$\sum_{k=0}^{n} M_{3k+2} = M_{3n+3} - 3,$$

$$\sum_{k=0}^{n} M_{3k} = M_{3n+1} + 2.$$

We can also use the numbers M_n to calculate the numbers G_{-n}.

Theorem 4.4. *Let n be a nonnegative integer. Then*

(1) $G_{-n} = G_{2n} - G_n M_n$, *and*
(2) $M_{-n} = \frac{1}{2}(M_n^2 - M_{2n})$.

Proof. We use the notation of Section 3.8. By Vieta's formula, $\alpha + \beta + \gamma = 1$, $\alpha\beta\gamma = 1$, $\alpha\beta + \beta\gamma + \gamma\alpha = 0$; also $A + B + C = G_0 = 0$. By the nature of their recurrence relations, the characteristic polynomial of M_n is the same as the characteristic polynomial of G_n: $x^3 - x^2 - 1 = 0$. Assume $M_n = C_1\alpha^n + C_2\beta^n + C_3\gamma^n$. Then from the initial conditions for the sequence M_n, we have

$$\begin{pmatrix} 1 & 1 & 1 \\ \alpha & \beta & \gamma \\ \alpha^2 & \beta^2 & \gamma^2 \end{pmatrix} \begin{pmatrix} C_1 \\ C_2 \\ C_3 \end{pmatrix} = \begin{pmatrix} 3 \\ 1 \\ 1 \end{pmatrix}.$$

Since the characteristic polynomial has no multiple roots, we have the unique solution

$$\begin{pmatrix} C_1 \\ C_2 \\ C_3 \end{pmatrix} = \begin{pmatrix} 1 \\ 1 \\ 1 \end{pmatrix},$$

so $M_n = \alpha^n + \beta^n + \gamma^n$. Then by calculating directly with explicit expressions,

$$\begin{aligned} G_{2n} - G_n M_n &= A\alpha^{2n} + B\beta^{2n} + C\gamma^{2n} \\ &\quad -(A\alpha^n + B\beta^n + C\gamma^n)(\alpha^n + \beta^n + \gamma^n) \\ &= -(A+B)(\alpha\beta)^n - (B+C)(\beta\gamma)^n - (C+A)(\gamma\alpha)^n \\ &= A\alpha^{-n} + B\beta^{-n} + C\gamma^{-n} \\ &= G_{-n}. \end{aligned}$$

This proves (1). Similarly, (2):

$$\begin{aligned} M_n^2 - M_{2n} &= (\alpha^n + \beta^n + \gamma^n)^2 - (\alpha^{2n} + \beta^{2n} + \gamma^{2n}) \\ &= 2((\alpha\beta)^n + (\beta\gamma)^n) + (\gamma^\alpha)^n) \\ &= 2(\alpha^{-n} + \beta^{-n} + \gamma^{-n}) \\ &= 2M_{-n}, \end{aligned}$$

as required. $\qquad\square$

We make the following conjectures.

Conjecture 4.1. *The only integers n such that $G_n = 0$ are $n = 0$, -1, -3, -8.*

Conjecture 4.2. *The only integers n such that $M_n = 0$ are $n = -1$, -11.*

By Theorem 4.4, the two conjectures above are equivalent to (1) $G_{2n} = G_n M_n$ if and only if $n = 0$, -1, -3, or -8, and (2) $M_{2n} = M_n^2$ if and only if $n = -1$ or -11.

Finally, we have the following generalization of the determinant identity in Section 3.5.

Theorem 4.5. *Suppose i, j, k, m, n are all nonnegative integers. Then*

$$\begin{vmatrix} F_i & F_{i+m} & F_{i+n} \\ F_j & F_{j+m} & F_{j+n} \\ F_k & F_{k+m} & F_{k+n} \end{vmatrix} = 0.$$

This identity remains valid if the Fibonacci numbers in the determinant are replaced by Lucas numbers.

4.2. Criteria for Fibonacci Numbers

It can be proven from the Binet formula that a positive integer x is a Fibonacci number if and only if at least one of $5x^2 + 4$ and $5x^2 - 4$ is a square number. In fact, if

$$x = \log_\phi \frac{F_n \sqrt{5} + \sqrt{5F_n^2 \pm 4}}{2}$$

is an integer, then the term under the second radical sign must be a square number.

There also some interesting and famous congruences involving odd prime moduli (the second congruence below is also valid for some composite moduli):

$$F_p \equiv (p \mid 5) \pmod{p},$$
$$L_p \equiv 1 \pmod{p}, \tag{4.4}$$
$$F_{p-(p \mid 5)} \equiv 0 \pmod{p},$$

where $(p \mid 5)$ is the Legendre symbol for quadratic residues,

$$(p \mid 5) = \begin{cases} 0 & \text{if } p = 5, \\ 1 & \text{if } p \equiv \pm 1 \pmod 5, \\ -1 & \text{if } p \equiv \pm 2 \pmod 5. \end{cases}$$

From (4.4) and Lucas's theorem, we can derive the following theorem.

Theorem 4.6. *Suppose F_n is the first Fibonacci number among the multiples of p, then $p \equiv (p \mid 5) \pmod n$. In particular, if $p = F_q$ is a Fibonacci prime, then q is prime, and $p \equiv (p \mid 5) \pmod q$.*

Proof. Suppose p is a prime number, n is the smallest positive integer such that p divides F_n. Suppose also that p divides F_m for some m such that n does not divide m. It follows from Theorem 3.4 that p divides $F_{\gcd(m,n)}$; but $\gcd(m, n) < n$, which in a contradiction. Therefore n divides m whenever p divides F_m, and the result follows from (4, 4). $\qquad\qquad\qquad\qquad\qquad\qquad\qquad\qquad\qquad\square$

In 1992, Zhi-Hong Sun and Zhi-Wei Sun proved (*Fibonacci numbers and Fermat's last theorem, Acta Arithmetica*, 60.4, 371–388) the following theorem (Fig. 4.3).

Theorem 4.7. *Let p be a prime number. Then we have the following properties:*

(1) *If $p > 5$ and $p \equiv 1 \pmod 4$, then $F_{(p-(p \mid 5))/2} \equiv 0 \pmod p$.*
(2) *If $p \equiv 3 \pmod 4$, then $L_{(p-(p \mid 5))/2} \equiv 0 \pmod p$.*

If we consider the case $p \equiv 1 \pmod 4$ it is obvious that $\frac{p-(p \mid 5)}{4}$ is an integer only when $p \equiv 1$ or $9 \pmod{20}$. The Sun brothers also proved that if $p \equiv 1$ or $9 \pmod{20}$ and $p = x^2 + 5y^2$ for some integers x and y, then:

$$p \text{ divides } F_{(p-1)/4} \text{ if and only if } 4 \text{ divides } xy. \qquad (4.5)$$

The second congruence in Theorem 4.7 shows that if some prime $p \equiv 3 \pmod 4$, then there must be Lucas numbers among its multiples. It can also happen that there are Lucas numbers among the multiples of a prime $p \equiv 1 \pmod 4$. For example, $L_7 = 29$, $L_{10} = 3 \times 41$. It is worth pointing out also that $\frac{p-(p \mid 5)}{2}$ is not necessarily the

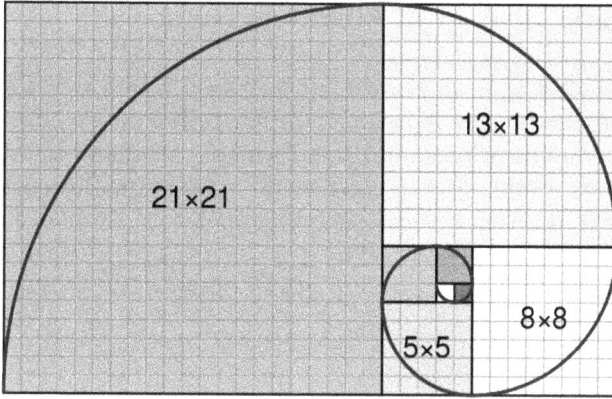

Figure 4.2.   The Fibonacci spiral

smallest index for which the above congruences hold. For example, $L_8 = 47$. As for (4.5), we point out also that under the same hypotheses ($p \equiv 1$ or $9 \pmod{20}$ and $p = x^2 + 5y^2$ for some integers x and y), p divides $L_{(p-1)/2}$ if and only if 4 does *not* divides xy.

Furthermore, we have the following theorem.

Theorem 4.8. *If $p \equiv 17$ or $23 \pmod{30}$ and $p = 3x^2 + 5y^2$ for some integers x and y, then we have the following properties*

(1) *p divides $F_{(p-1)/6}$ if and only if 6 divides xy.*
(2) *p divides $L_{(p-1)/6}$ if and only if 6 does not divide xy.*

Finally, we present two conjectures.

Conjecture 4.3. *Suppose p is a prime number and n is the smallest positive integer such that p divides F_n. Then*

$$\frac{p - (p \mid 5)}{n} \equiv \begin{cases} 0 \pmod{2} & \text{if } p \neq 5 \text{ and } p \equiv 1 \pmod{4}, \\ 1 \pmod{2} & \text{if } p = 5 \text{ or } p \equiv 3 \pmod{4}. \end{cases}$$

Conjecture 4.4. *Suppose p and n is the smallest positive integer such that p divides L_n. Then*

$$\frac{p - (p \mid 5)}{2n} \equiv \begin{cases} 0 \pmod{2} & \text{if } p \equiv 1 \pmod{4}, \\ 1 \pmod{2} & \text{if } p \equiv 3 \pmod{4}. \end{cases}$$

4.3. Prime Divisors of Fibonacci Numbers

The congruence (4.4) in the previous section shows that every prime number divides some Fibonacci number. In particular, if p and $p+2$ are a pair of twin primes, and $p \equiv 7 \pmod{10}$, then $p(p+2)$ divides F_{p+1}. For example, $F_{18} = 2584 = 2^3 \times 17 \times 19$. Moreover, if n is the smallest positive integer such that $p(p+2)$ divides F_n, then it follows from Conjecture 4.1 that if $p \equiv 7 \pmod{10}$, then $n = p+1$. More generally, Conjecture 4.1 shows that $n = \mathrm{lcm}(p - (p \mid 5), q - (q \mid 5))$. So

$$
n = \begin{cases}
p+1 & \text{if } p \equiv 7 \pmod{10}, \\
(p^2 - 1)/2 & \text{if } p \equiv 9 \pmod{10}, \\
(p-1)(p+3)/2 & \text{if } p \equiv 11 \pmod{20}, \\
(p-1)(p+3)/4 & \text{if } p \equiv 1 \pmod{20}.
\end{cases}
$$

We now prove (4.4).

Proof of (4.4). If $p = 2$ or 5 there is nothing to prove. Suppose $p \neq 2$ or 5. By Binet's formula,

$$
2^{n-1} F_n = n + \binom{n}{3} 5 + \binom{n}{5} 5^2 + \cdots, \tag{4.6}
$$

where the final term is $5^{(n-1)/2}$ if n is odd, and $n \cdot 5^{(n-2)/2}$ if n is even. Set $n = p$. Then by Fermat's little theorem and Euler's quadratic residue criterion,

$$
2^{p-1} \equiv 1 \pmod{p},
$$
$$
5^{(p-1)/2} \equiv (5 \mid p) \pmod{p}.
$$

Since each term in the sum (4.6) except the last is a multiple of p, this gives

$$
F_p \equiv (5 \mid p) \pmod{p}.
$$

Next, observe that from Cassini's formula,

$$
F_{p-1} F_{p+1} \equiv 0 \pmod{p}.
$$

Pierre de Fermat.

Figure 4.3.   French mathematician Pierre de Fermat

Next, since $\gcd(p-1, p+1) = 2$, then from Lucas's identity (3.6),

$$\gcd(F_{p-1}, F_{p+1}) = F_2 = 1.$$

It follows that p divides only one of F_{p-1} and F_{p+1}. We would like to sort of which of them it is; set $n = p+1$ in (4.6). Then apart from the first and last term, every other term is still divisible by p, so

$$2^p F_{p+1} \equiv 1 + (5 \mid p) \pmod{p}.$$

Therefore $F_{p+1} \equiv 0 \pmod{p}$ if $(5 \mid p) = -1$, and $F_{p-1} \equiv 0 \pmod{p}$ if $(5 \mid p) = 1$. This completes the proof of (4.4).   □

For the Lucas numbers, we have

$$L_{p-(p \mid 5)} \equiv 2 (p \mid 5) \pmod{p},$$

$$L_{p+(p \mid 5)} \equiv 3 (p \mid 5) \pmod{p}, \tag{4.7}$$

for all primes $p \neq 5$. The first of these congruences can be found in [8].

Next, we give a different proof of Lucas's theorem. First, from the recursions involved and (4.2), it is easy to see that

(1) $\gcd(F_n, F_{n+1}) = \gcd(L_n, L_{n+1}) = 1$,
(2) F_n and L_n have the same parity, and
(3) $\gcd(F_n, L_n) = 1$ or 2.

We prove first that for every integer r, F_n divides F_{nr}. Suppose first that r is positive. For $r = 1$ or 2, it is obvious. Suppose $r > 2$ and the result holds for $r - 1$. By (4.1),

$$2F_{rn} = F_n L_{(r-1)n} + F_{(r-1)n} L_n.$$

If F_n is odd, it follows immediately from the induction hypothesis that F_n divides F_{rn}; otherwise, if F_n is even, it follows from the previous analysis and the induction hypothesis that L_n, $L_{(r-1)n}$, and $F_{(r-1)n}$ are all even, so again F_n divides F_{rn}. Finally, since $F_{-n} = (-1)^{n-1} F_n$, the result extends also to all integers n.

We now prove (3.9). Write $d = \gcd(m, n)$. By Euclidean division there are integers r and s such that

$$d = rm + sn.$$

Therefore by (4.1),

$$2F_d = F_{rm} L_{sn} + F_{sn} L_{rm}.$$

Suppose $D = \gcd(F_m, F_n)$. Then by the argument in the preceding paragraph, D divides F_{rm} and F_{sn}, so also D divides $2F_d$. If D is odd, then automatically D divides F_d. If D is even, then both F_m and F_n are even, so all of F_{rm}, F_{sn}, L_{rm}, L_{sn} are even; again D divides F_d. Finally, F_d divides both F_n and F_m, so we have in the other direction that F_d divides D, therefore $D = F_d$, or $\gcd(F_m, F_n) = F_{\gcd(m,n)}$. □

4.4.   Fibonacci Congruences

It is well known that the Fibonacci numbers and Lucas numbers satisfy the following congruences (see [8]):

$$F_p \equiv (p \mid 5) \pmod{p} \quad \text{and} \quad F_{p-(p \mid 5)} \equiv 0 \pmod{p}, \quad (4.8)$$

$$L_p \equiv 1 \pmod{p} \quad \text{and} \quad 2L_{p-(p \mid 5)} \equiv 2(p \mid 5) \pmod{p}. \quad (4.9)$$

We note that such formulas can be generalized via the addition rule; suppose $p \neq 5$, k is any integer; then

$$F_{p-k(p\,|\,5)} \equiv \begin{cases} -F_{k-1} \pmod{p} & \text{if } k \text{ is odd,} \\ (p \mid 5) F_{k-1} \pmod{p} & \text{if } k \text{ is even,} \end{cases} \tag{4.10}$$

$$L_{p-k(p\,|\,5)} \equiv \begin{cases} (p \mid 5) L_{k-1} \pmod{p} & \text{if } k \text{ is odd,} \\ -L_{k-1} \pmod{p} & \text{if } k \text{ is even.} \end{cases} \tag{4.11}$$

When $k = 0$ or 1, the latter congruences are identical with (4.8) and (4.9), respectively. If $p = 5$, then $k = 1$ is the only odd number for which (4.10) holds, and $k = 0$ the only number for which (4.11) holds.

Also for any prime $p \neq 5$, we have (see [8]):

$$\begin{aligned} F_{n+p-(p\,|\,5)} &\equiv (p \mid 5) F_n \pmod{p}, \\ L_{n+p-(p\,|\,5)} &\equiv (p \mid 5) L_n \pmod{p}. \end{aligned} \tag{4.12}$$

We can improve these congruences for all $p \neq 5$, $p < 50$ to the following identities:

$$F_{n+3} + F_n = 2F_{n+2},$$

$$F_{n+4} + F_n = 3F_{n+2},$$

$$F_{n+8} + F_n = 7F_{n+4},$$

$$F_{n+10} - F_n = 11F_{n+5},$$

$$F_{n+14} + F_n = 13(F_{n+8} + F_{n+6}),$$

$$F_{n+18} + F_n = 17(F_{n+12} + F_{n+6}),$$

$$F_{n+18} - F_n = 19(F_{n+12} - F_{n+6}),$$

$$F_{n+24} + F_n = 23((F_{n+17} - F_{n+7}) + (F_{n+14} + F_{n+10}))$$
$$= 23(2 \cdot (F_{n+16} + F_{n+8})),$$

$$F_{n+28} - F_n = 29(F_{n+22} - F_{n+16}),$$

$$F_{n+30} - F_n = 31((F_{n+23} + F_{n+7}) - (F_{n+17} + F_{n+13})),$$
$$= 32(4 \cdot (F_{n+20} - F_{n+10})),$$

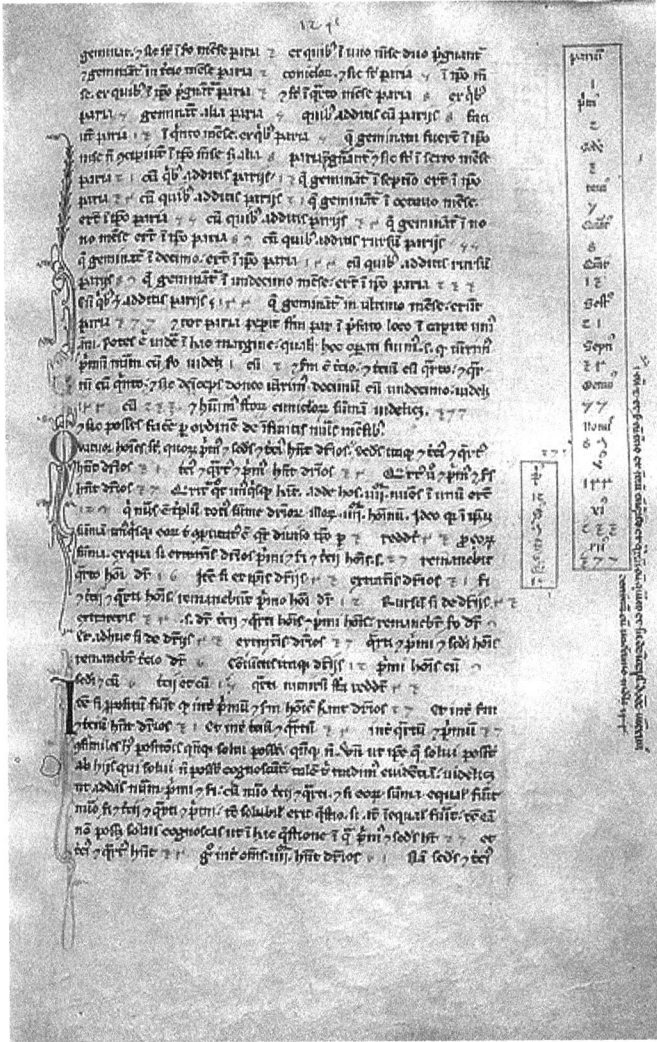

Figure 4.4. First page of *Liber Abaci*, by Fibonacci

$$F_{n+38} + F_n = 37((F_{n+23} - F_{n+15}) + (F_{n+27} + F_{n+11})),$$
$$F_{n+40} - F_n = 41((F_{n+28} - F_{n+12}) + (F_{n+32} - F_{n+8})),$$
$$F_{n+44} + F_n = 43((F_{n+36} + F_{n+8}) + (F_{n+31} - F_{n+13}) + 2F_{n+22}),$$
$$F_{n+48} + F_n = 47((F_{n+40} + F_{n+8}) - F_{n+24}).$$

Each of the above recurrences can be proved by induction. Since each positive integer admits a presentation as a sum of nonadjacent Lucas numbers, and the Lucas numbers satisfy

$$L_n = F_{n+1} + F_{n-1},$$

it is easy to see from the usual recurrences and congruence (4.12) that for any prime $p \neq 5$, the same or similar identities hold for the Lucas numbers, and more generally for terms in any Lucas sequence (see Section 4.9).

4.5. A More General Congruence

In 1963, J.A. Maxwell identified (*Fibonacci Quarterly*, 1:1, Feb., 75) congruences:

$$2^n F_{n+1} + 2^{n+1} F_n \equiv 1 \pmod 5,$$
$$3^n F_{n+1} + 3^{n+1} F_n \equiv 1 \pmod{11},$$
$$5^n F_{n+1} + 5^{n+1} F_n \equiv 1 \pmod{29},$$

where n is any nonnegative integer. These congruences are all special cases of the following generalization: if p is any prime number and n a nonnegative integer, then

$$p^n F_{n+1} + p^{n+1} F_n \equiv 1 \pmod{p^2 + p - 1},$$

in turn a special case of an even more general theorem.

First, we define the *generalized Fibonacci numbers* $F_{(n;a,b)}$ with a and b arbitrary fixed integers by

$$\begin{cases} F_{(0;a,b)} = b - a, \\ F_{(1;a,b)} = a, \\ F_{(2;a,b)} = b, \\ F_{(n;a,b)} = F_{(n-1;a,b)} + F_{(n-2;a,b)} \quad (n \geq 3). \end{cases}$$

For fixed a, b we shall write $J_n = F_{(n;a,b)}$ for brevity. When $a = b = 1$, $J_n = F_n$, and when $a = 1$, $b = 3$, $J_n = L_n$. In general, the terms of the sequence $(J_n)_{n \geq 1}$ are

$$(b-a), a, b, a+b, a+2b, 2a+3b, 3a+5b, \ldots, aF_{n-2} + bF_{n-1}, \ldots,$$

where the general expression $J_n = aF_{n-2} + bF_{n-1}$ for $n \geq 3$ can be obtained by induction. We also have the following version of the sixth equation in (4.1):

$$J_{m+n} = J_{m+1}F_n + J_nF_{n-1}.$$

Thomas Koshy proved the next two theorems in 1999.

Theorem 4.9 If m is any integer and n a nonnegative integer, then

$$J_{n+1}m^n + J_nm^{n+1} \equiv a(1-m) + bm \pmod{m^2 + m - 1}.$$

Proof. We induct on n. When $n = 0$, this is simply an identity:

$$J_1m^0 + J_0m^1 = a + (b-a)m = a(1-m) + bm.$$

When $n = 1$, this is

$$J_2m^1 + J_1m^2 = bm + am^2$$

$$\equiv a(1-m) + bm \pmod{m^2 + m - 1}.$$

Suppose then that $n \geq 1$ and the result holds for all $0 \leq k \leq n$. Then

$$J_nm^{n-1} + J_{n-1}m^n \equiv a(1-m) + bm \pmod{m^2 + m - 1},$$
$$J_{n+1}m^n + J_nm^{n+1} \equiv a(1-m) + bm \pmod{m^2 + m - 1},$$

so

$$J_{n+2}m^{n+1} + J_{n+1}m^{n+2} = (J_n + J_{n+1})m^{n+1} + (J_{n-1} + J_n)m^{n+2}$$
$$= m(J_{n+1}m^n + J_nm^{n+1})$$
$$\quad + m^2(J_nm^{n-1} + J_{n-1}m^n)$$
$$\equiv m\,(a(1-m) + bm) + m^2\,(a(1-m) + bm)$$
$$\equiv (m^2 + m)\,(a(1-m) + bm)$$
$$\equiv a(1-m) + bm \pmod{m^2 + m - 1},$$

as required.

Corollary 4.1. *Since the Fibonacci numbers and the Lucas numbers are both special cases of generalized Fibonacci numbers, therefore for all integers m and all nonnegative integers n,*

$$F_{n+1}m^n + F_n m^{n+1} \equiv 1 \pmod{m^2 + m - 1},$$
$$L_{n+1}m^n + L_n m^{n+1} \equiv 1 + 2m \pmod{m^2 + m - 1}.$$

Theorem 4.10. *If $n \geq 0$, $m \geq 2$, then*

$$F_{n-1} - mF_n \equiv (-1)^n m^n \pmod{m^2 + m - 1}. \tag{4.13}$$

Proof. We induct again on n. When $n = 0$ and $n = 1$, we have the straightforward identities:

$$F_{-1} - mF_0 = 1 = (-1)^0 m^0,$$
$$F_0 - mF_1 = -m = (-1)^1 m^1,$$

respectively. Let $n \geq 1$ and suppose the result is valid for n and $n-1$; then

$$
\begin{aligned}
F_n - mF_{n+1} &= (F_{n-2} + F_{n-1}) - m(F_{n-1} + F_n) \\
&= (F_{n-2} - mF_{n-1}) + (F_{n-1} - mF_n) \\
&\equiv (-1)^{n-1} m^{n-1} + (-1)^n m^n \\
&\equiv (-1)^{n-1} m^{n-1}(1 - m) \\
&\equiv (-1)^{n-1} m^{n-1} m^2 \\
&\equiv (-1)^{n+1} m^{n+1} \pmod{m^2 + m - 1}.
\end{aligned}
$$

This completes the induction. $\square$

Unfortunately, there does not seem to be any good analogue to Theorem 4.10 for the Lucas numbers. We do, however, have the following corollaries.

Corollary 4.2. *Every $L_n \equiv (-1)^n 2^{n+1} \equiv 2 \cdot 3^n \pmod 5$.*

This follows directly from Theorem 4.10:

$$
\begin{aligned}
L_n &= 2F_{n+1} - F_n \\
&= -(F_n - 2F_{n+1}) \\
&\equiv (-1)^n 2^{n+1} \pmod{2^2 + 2 - 1 = 5}.
\end{aligned}
$$
$\square$

Corollary 4.3. *For nonnegative n,*

- $L_{4n} \equiv 2 \pmod 5$,
- $L_{4n+1} \equiv 1 \pmod 5$,
- $L_{4n+2} \equiv 3 \pmod 5$,
- $L_{4n+3} \equiv 4 \pmod 5$.

4.6.   Narayana Sequence Congruences

We have introduced the Narayana cow sequence already in Section 3.2, and in Section 4.1 gave the expression $G_{-n} = G_{2n} - G_n M_n$ for its negative indexed terms. We observe first that using the six identities with a coefficient ± 1 on the right-hand side among the 15 identities given in (3.8), we obtain six more expression for G_{-n} in which the numbers M_n play no role:

$$G_{-n} = G_n G_{n-4} - G_{n-1} G_{n-3},$$
$$G_{-n} = G_{n-1} G_{n-4} - G_{n-2} G_{n-3},$$
$$G_{-n} = G_{n-3} G_{n-4} - G_{n-1} G_{n-6},$$
$$G_{-n} = G_{n-4} G_{n-8} - G_{n-3} G_{n-9},$$
$$G_{-n} = G_{n-6} G_{n-8} - G_{n-3} G_{n-11},$$
$$G_{-n} = G_{n-4} G_{n-11} - G_{n-6} G_{n-9}.$$

It is also not difficult to prove two analogues of Cassini's identity using the Narayana recurrence:

$$G_{-n-1} = G_{n-1} G_{n+1} - G_n^2,$$
$$G_{-n-3} = G_{n-1} G_{n+2} - G_n G_{n+1}.$$

We give below several identities and congruences concerning the numbers G_n and M_n that were discovered and proved in the course of our weekly graduate number theory seminar in the autumn semester of 2020 through the contributions of Xiaoyu Wang, Yong Chen, and Zhongyan Chen. In particular, we have an analogue to Vajda's identity. First, we will need a few lemmas.

Lemma 4.1. *We have the following identities for the numbers F_n and G_n:*

$$F_{pn} = \sum_{k=0}^{p} \binom{p}{k} F_{p(n-1)-k},$$

$$G_{pn} = \sum_{k=0}^{p} \binom{p}{k} G_{p(n-1)-2k}.$$

This lemma can be proved by repeated applications of the recurrences involved and the binomial coefficient identity

$$\binom{n}{k} + \binom{n}{k+1} = \binom{n+1}{k+1}.$$

From Lemma 4.1, we can prove the following theorem, making use of the familiar fact that the prime p divides $\binom{p}{k}$ for every $1 \le k \le p-1$.

Theorem 4.11. *For all integers n and primes p,*

$$G_{pn} \equiv G_{p(n-1)} + G_{p(n-3)} \pmod{p}.$$

From this theorem, and induction on n, we also have the following theorem.

Theorem 4.12. *For p prime and all integers n,*

$$F_{pn} \equiv F_p F_n \pmod{p},$$
$$G_{pn} \equiv G_p G_n + (G_{2p} - G_p) G_{n-1} \pmod{p}.$$

The next theorem presents some congruences for the numbers M_n.

Theorem 4.13. *For any prime p,*

$$M_p \equiv 1 \pmod{p},$$
$$M_{2p} \equiv 1 \pmod{p},$$
$$M_{-p} \equiv 0 \pmod{p}.$$

Proof. By Vieta's formula, $\alpha+\beta+\gamma = 1$, $\alpha\beta\gamma = 1$, $\alpha\beta+\beta\gamma+\gamma\alpha = 0$; therefore

$$(\alpha + \beta + \gamma)^p = \sum_{\substack{0 \leq i,j,k \\ i+j+k=p}} \frac{p!}{i!j!k!} \alpha^i \beta^j \gamma^k$$

$$= \alpha^p + \beta^p + \gamma^p + p \sum_{\substack{0 \leq i,j,k < p \\ i+j+k=p}} \frac{(p-1)!}{i!j!k!}$$

$$= 1.$$

Since the coefficients in the final sum are integers, it follows from the basic theory of symmetric polynomials that this can be written as a polynomial in coefficients $\alpha + \beta + \gamma$, $\alpha\beta\gamma$, and $\alpha\beta + \beta\gamma + \gamma\alpha$. Since everything in the sum is an integer, the sum too must be an integer. It follows that

$$M_p \equiv (\alpha + \beta + \gamma)^p \equiv 1 \pmod{p}.$$

The proof that $M_{2p} \equiv 1 \pmod{p}$ is similar. We prove next that $M_{-p} \equiv 0 \pmod{p}$. Note that

$$M_{-p} = \alpha^{-p} + \beta^{-p} + \gamma^{-p}$$
$$= (\alpha\beta)^p + (\beta\gamma)^p + (\gamma\alpha)^p$$
$$= (\alpha\beta + \beta\gamma + \gamma\alpha)^p + pf(\alpha, \beta, \gamma)$$
$$= pf(\alpha, \beta, \gamma),$$

where f is a polynomial with integer coefficients. Therefore $M_{-p} \equiv 0 \pmod{p}$. $\qquad\square$

From Theorem 4.4, the second formula in Theorem 4.12, and Theorem 4.13, we get the following theorem.

Theorem 4.14. *If p is prime and n is any integer, then*

$$G_{pn} \equiv G_p G_n + G_{-p} G_{n-1} \pmod{p}.$$

On the other hand, it is easy to prove by induction that $M_n = G_n + 3G_{n-2}$. Noting that the period of G_n modulo 3 is 8, then by induction on k, we also have the following result.

Theorem 4.15. *If $k \geq 1$ and $n \equiv -1, 1$, or $2 \pmod{3^{k-1}8}$, then*

$$M_n \equiv G_n \pmod{3^{k+1}}.$$

In order to prove the Vajda identity analogue for the Narayana cow sequence, we need a corresponding addition rule:

$$G_{m+n} = G_{m+1}G_{n_1} + G_{m-1}G_{n-1} - G_{m-2}G_{n-2}.$$

This can be proved by induction on n for fixed m. Then we obtain the following analogue to the Vajda identity:

$$G_{n+i}G_{n+j} - G_nG_{n+i+j} = G_iG_jG_{-n-2} - (G_iG_{j-1} + G_{i-1}G_j)G_{-n-1}$$
$$+ G_{i-1}G_{j-1}G_{-n}.$$

Using the method of generating functions, we can also derive the following identities:

$$G_{3n} = \sum_{k=0}^{n} \binom{n}{k} G_{2k},$$

$$(-1)^n G_{2n} = \sum_{k=0}^{n} \binom{n}{k} (-1)^k G_{2k},$$

$$G_n = \sum_{k=0}^{n} \binom{n}{k} G_{-2k}.$$

In general, if we consider the sequence defined by $H_0 = 0$, $H_1 = \cdots = H_t = 1$, $H_n = H_{n-1} + H_{n-t}$ $(n \geq t \geq 2)$, we have

$$H_{tn} = \sum_{k=0}^{n} \binom{n}{k} H_{(t-1)k},$$

$$(-1)^n H_{(t-1)n} = \sum_{k=0}^{n} \binom{n}{k} (-1)^k H_{(t-1)k},$$

$$H_n = \sum_{k=0}^{n} \binom{n}{k} H_{-(t-1)k}.$$

4.7. Pythagorean Triples

A *Pythagorean triple* (x, y, z) is any integer solution to the quadratic equation

$$x^2 + y^2 = z^2. \tag{4.14}$$

As we have mentioned at the start of this book, it has long been known in China that the smallest Pythagorean triple is $(3, 4, 5)$ (Figs. 4.5 and 4.6). Clay tablets left behind by the Babylonians suggest that they may have known about the exist of this Pythagorean triple (and the Pythagorean theorem) even earlier. In any case, the ancient Greeks were the first to write down a rule for the generation of Pythagorean triples, namely

$$\begin{cases} x = 2n + 1, \\ y = 2n^2 + 2n, \\ z = 2n^2 + 2n + 1. \end{cases}$$

This result was attributed to Pythagoras himself by Proclus (ca. 410–485), the last major philosopher of the ancient Greek world.

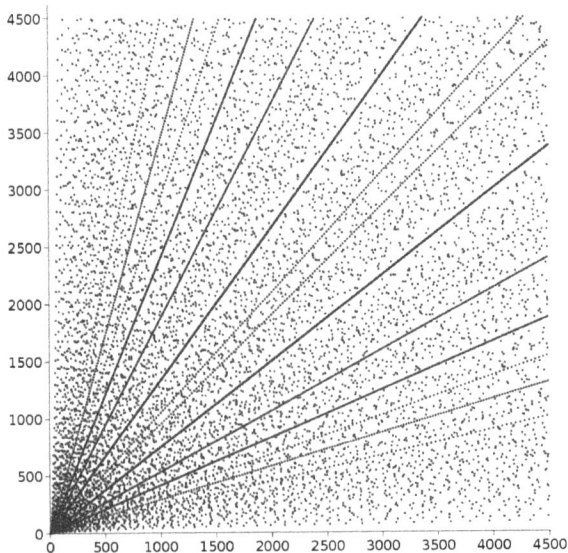

Figure 4.5.   Scatter plot of Pythagorean triplets

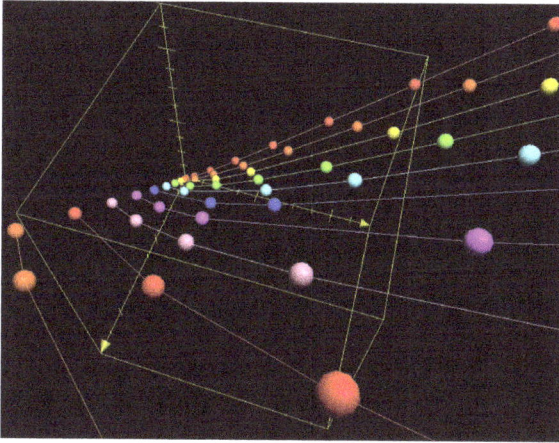

Figure 4.6. Scatter plot of Pythagorean triplets

Pythagoras was apparently particularly interested in right triangles such that the hypotenuse and one of the remaining sides differ by 1, for example $(3, 4, 5)$ $(n = 1)$, $(5, 12, 13)$ $(n = 2)$, $(7, 24, 25)$ $(n = 3)$. Another triple of small integer solutions is $(8, 15, 17)$, which does not follow the above rule. Proclus believed that it was Plato who discovered this triple. In particular, Plato studied triples of the form

$$\begin{cases} x = 2n, \\ y = n^2 - 1, \\ z = n^2 + 1. \end{cases}$$

When $n > 1$ is even, this always gives a Pythagorean triple, corresponding to a right triangle with hypotenuse longer by 2 than one of the remaining sides.

Euclid determined all irreducible solutions to this equation in his book *The Elements*, but as usual for that book, he did not mention the source of this result. He show that all Pythagorean triples with $\gcd(x, y) = 1$ have the form

$$\begin{cases} x = 2ab, \\ y = a^2 - b^2, \\ z = a^2 + b^2, \end{cases}$$

where $0 < a < b$, $\gcd(a, b) = 1$, and $a + b$ is odd. This proposition is called Euclid's formula, and its proof depends on the following observation: if a square number is equal to the product of two relatively prime positive integers, then the two numbers in the product must also be squares. Then if we write the quadratic equation (4.14) as

$$\left(\frac{x}{z}\right)^2 = \frac{z+y}{2} \cdot \frac{z-y}{2},$$

it follows from $\gcd(\frac{z+y}{2}, \frac{z-y}{2}) = 1$ that $\frac{z+y}{2} = a^2$ and $\frac{z-y}{2} = b^2$ for some integers a, b, from which every solution can be determined. The choice $a = n$, $b = 1$ gives Plato's solutions, and the choice $a = n+1$, $b = n$ gives Pythagoras's solutions. It is also possible to prove that for any prime number $p \equiv \pm 1 \pmod 8$, there exist infinitely many Pythagorean triples such that the difference between the lengths of the two legs is p. Using Euclid's formula, it is possible to generate every Pythagorean triple: either directly as a solution, or as an integer multiple of a solution.

Returning to the Fibonacci numbers, starting from $F_5 = 5$, every Fibonacci number with odd index is the largest number in some Pythagorean triple, equivalently the length of the hypotenuse of some right triangle with integer length sides; for example, $(3, 4, F_5 = 5)$, $(5, 12, F_7 = 13)$, $(16, 30, F_9 = 34)$. In fact, if you set

$$\begin{cases} a_n = F_{2n-1}, \\ b_n = 2F_n F_{n-1}, \\ c_n = F_n^2 - F_{n-1}^2, \end{cases}$$

then

$$a_n^2 = b_n^2 + c_n^2.$$

The Pythagorean solutions can also be regarded as a composition of four consecutive Fibonacci numbers. For $n \geq 1$, set

$$\begin{cases} e_n = F_n F_{n+3}, \\ f_n = 2F_{n+1} F_{n+2}, \\ g_n = F_{n+1}^2 + F_{n+2}^2. \end{cases}$$

Then again

$$g_n^2 = e_n^2 + f_n^2.$$

It follows immediately that the Pythagorean solutions can also be written in terms of four consecutive Lucas numbers. If

$$\begin{cases} h_n = L_n L_{n+3}, \\ i_n = 2L_{n+1} L_{n+2}, \\ j_n = L_{n+1}^2 + L_{n+2}^2, \end{cases}$$

then

$$j_n^2 = h_n^2 + i_n^2.$$

All of the above is easily verified using the relevant recurrences.

4.8. Diophantine m-tuples

In his book *Arithmetica*, the Greek mathematician Diophantus (ca. 246–330) put forward a problem that has since come to be called the problem of Diophantine m-tuples (Fig. 4.7). Diophantus observed that quadruple $\{\frac{1}{16}, \frac{33}{16}, \frac{17}{4}, \frac{105}{16}\}$ of rational numbers has the property that the product of any two of them is one less than a square (rational) number. Many centuries later, Fermat found the first quadruple of integers with this property: $\{1, 3, 8, 120\}$. Indeed,

$$1 \times 3 + 1 = 2^2, \qquad 1 \times 8 + 1 = 3^2, \qquad 1 \times 120 + 1 = 11^2,$$
$$3 \times 8 + 1 = 5^2, \qquad 3 \times 120 + 1 = 19^2, \qquad 8 \times 120 + 1 = 31^2.$$

A *Diophantine m-tuple* is an m-tuple $\{a_1, \ldots, a_m\}$ of distinct integers with the property that $a_j a_k + 1$ is a perfect square for all $1 \le j < k \le m$. An m-tuple of positive rational numbers with the same property is a *rational Diophantine m-tuple*.

Euler discovered an infinite class of Diophantine 4-tuples of the form

$$\{a, b, a + b + 2r, 4r(a + r)(b + r)\},$$

where $ab + 1 = r^2$. For example, with $r = 3$ and $a = 1$, $b = 8$, we get the Diophantine 4-tuple $\{1, 8, 15, 528\}$; with $a = 2, b = 4$, we get

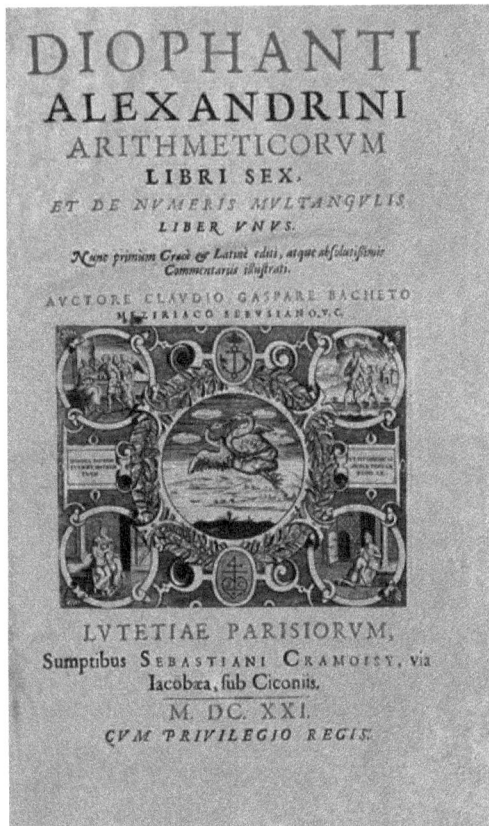

Figure 4.7.   Cover of the first Latin edition (1621) of *Arithmetica*, by Diophantus

the Diophantine 4-tuple $\{2, 4, 12, 300\}$. Euler also discovered that the rational number $\frac{777480}{8288641}$ can be appended to Fermat's Diophantine 4-tuple to obtain a rational Diophantine 5-tuple. In 1999, the British mathematician Alan Baker (1939–2018) discovered the first rational Diophantine 6-tuple:

$$\left\{ \frac{11}{192}, \frac{35}{192}, \frac{155}{27}, \frac{512}{27}, \frac{1235}{48}, \frac{180873}{16} \right\}.$$

For a long time, it was a folkloric conjecture that there do not exist any Diophantine 5-tuples. This was proved in 2019 by Bo He, Alain Togbé and Volker Ziegler (*Trans. Amer. Math. Soc.* **371** (2019), 6665–6709), building upon the cumulative results discussed below.

First, in 1969, the British mathematician Alan Baker (1939–2018) proved in collaboration with his teacher Harold Davenport (1907–1969) that the only d such that $\{1, 3, 8, d\}$ is a Diophantine 4-tuple is $d = 120$, using the theory of linear forms in the logarithms of algebraic numbers that Baker had pioneered. It follows that Fermat's Diophantine 4-tuple cannot be extended to a Diophantine 5-tuple. In 1998, the Croatian mathematician Andrej Dujella (1966–2012) improved this by proving that the pair $\{1, 3\}$ can be extended to a Diophantine 4-tuple in infinitely many ways, but not to any Diophantine 5-tuple. Dujella also obtained the best results to date on Diophantine m-tuples in general. He proved in 2004 that there are no Diophantine m-tuples for $m \geq 6$, and at most finitely many Diophantine 5-tuples.

On the other hand, Artin *et al.* proved in 1979 that every Diophantine 3-tuple can be extended to a Diophantine 4-tuple. Their approach is as follows: suppose $ab+1 = r^2$, $ac+1 = s^2$, $bc+1 = t^2$, and set $d_+ = a + b + c + 2abc + 2rst$. Then $\{a, b, c, d_+\}$ is a Diophantine 4-tuple. Dujella proposed the following conjecture concerning this construction.

Conjecture 4.4. *If $\{a, b, c, d\}$ is a Diophantine 4-tuple and $d > max(a, b, c)$, then $d = d_+$.*

Although this conjecture immediately implies that there are no Diophantine 5-tuples, it was not used in its proof of this result, and remains open.

If the terms $a_j a_k + 1$ are replaced in the definition of Diophantine m-tuples with $a_j a_k + s$ for some fixed positive integer s, then the situation is not the same. If we use the notation $P_s\{a_1, \ldots, a_m\}$ to indicate that the integers $a_1, \ldots, a_m$ have this property, then people have found already the following:

$$P_9\{1, 7, 40, 216\},\ P_{-7}\{1, 8, 11, 16\},\ P_{256}\{1, 33, 105, 320, 18240\},$$
$$P_{2985984}\{99, 315, 9920, 32768, 44460, 19534284\}.$$

It is an open question whether there exist bounds on the numbers s and m in this formulation.

In 1977, Hoggatt and Bergum (*Fibonacci Quarterly*, 15, 323–330) used Fibonacci numbers to construct a the Diophantine 4-tuples

$$\{F_{2n}, F_{2n+2}, F_{2n+4}, 4F_{2n+1}F_{2n+2}F_{2n+3}\}$$

for $n \geq 1$. For $n = 1$, $n = 2$, this gives the Diophantine 4-tuples $\{1, 2, 8, 120\}$ and $\{3, 8, 21, 2080\}$, respectively. This construction depends on the following easy consequences of Catalan's identity:

$$F_{2n}F_{2n+2} + 1 = F_{2n+1}^2,$$
$$F_{2n}F_{2n+4} + 1 = F_{2n+2}^2,$$
$$F_{2n+2}F_{2n+4} + 1 = F_{2n+3}^2. \tag{4.15}$$

We want a positive integer x such that $F_{2n}x+1$, $F_{2n+2}x+1$, $F_{2n+4}x+1$ are all perfect squares. By (4.15) and the Fibonacci recurrence, we have

$$1 = F_{2n+1}^2 - F_{2n}F_{2n+2} = F_{2n+1}F_{2n+2} - F_{2n}F_{2n3},$$

so

$$4F_{2n}F_{2n+1}F_{2n+2}F_{2n+3} + 1 = (2F_{2n+1}F_{2n+2} - 1)^2.$$

If we replace n with $n+1$ in (4.15), then we also have

$$1 = F_{2n+3}^2 - F_{2n+1}F_{2n+4} - F_{2n+3}F_{2n+4} = F_{2n+1}F_{2n+4} - F_{2n+2}F_{2n3},$$

which gives

$$4F_{2n+1}F_{2n+2}F_{2n+3}F_{2n+4} + 1 = (2F_{2n+2}F_{2n+3} + 1)^2.$$

Finally, if we multiply both sides of Cassini's identity $F_{2n+2}^2 = F_{2n+1}F_{2n+3} - 1$ by $4F_{2n+2}^2$, we get

$$4F_{2n+1}F_{2n+1}F_{2n+2}^2F_{2n+3} + 1 = (2F_{2n+2}^2 + 1)^2.$$

We conclude $x = 4F_{2n+1}F_{2n_2}F_{2n+3}$ is a solution. Hoggatt and Bergum further conjectured that this is the only solution. This conjecture was proved in 1999 by Dujella (*Proc. Amer. Math. Soc.* 127, 1999–2005).

There do not seem to be similar constructions using the Lucas numbers; even in the more unrestricted case $s \neq 1$, it is difficult to find examples. We can, however, prove the following near-miss, so to speak: for $n \geq 0$, the 4-tuplet

$$\{L_{2n+1}, L_{2n+3}, L_{2n+5}, 4L_{2n+2}L_{2n+3}L_{2n+4}\}$$

satisfies the identities

$$L_{2n+1}L_{2n+1} + 5 = L_{2n+2}^2,$$
$$L_{2n+1}L_{2n+5} + 5 = L_{2n+3}^2,$$
$$L_{2n+3}L_{2n+5} + 5 = L_{2n+4}^2,$$
$$L_{2n+1}(4L_{2n+2}L_{2n+3}L_{2n+4}) + 25 = (2L_{2n+2}L_{2n+3} - 5)^2,$$
$$L_{2n+3}(4L_{2n+2}L_{2n+3}L_{2n+4}) + 25 = (2L_{2n+3}^2 + 5)^2,$$
$$L_{2n+5}(4L_{2n+2}L_{2n+3}L_{2n+4}) + 25 = (2L_{2n+3}L_{2n+4} - 5)^2.$$

Similarly, for $n \geq 1$, the 4-tuplet

$$\{L_{2n}, L_{2n+2}, L_{2n+4}, 4L_{2n+1}L_{2n+2}L_{2n+3}\}$$

satisfies

$$L_{2n}L_{2n+2} - 5 = L_{2n+1}^2,$$
$$L_{2n}L_{2n+4} - 5 = L_{2n+2}^2,$$
$$L_{2n+2}L_{2n+4} - 5 = L_{2n+3}^2,$$
$$L_{2n}(4L_{2n+1}L_{2n+2}L_{2n+3}) + 25 = (2L_{2n+1}L_{2n+2} - 5)^2,$$
$$L_{2n+2}(4L_{2n+1}L_{2n+2}L_{2n+3}) + 25 = (2L_{2n+2}^2 + 5)^2,$$
$$L_{2n+4}(4L_{2n+1}L_{2n+2}L_{2n+3}) + 25 = (2L_{2n+2}L_{2n+3} - 5)^2.$$

4.9. Generating Functions

The generating function for the Fibonacci numbers is the power series

$$s(x) = \sum_{k=0}^{\infty} F_k x^k,$$

which converges for all $|x| < \frac{\sqrt{5}-1}{2} = 0.618\ldots$ to

$$s(x) = \frac{x}{1-x-x^2}.$$

This is because

$$s(x) = F_0 + F_1 x + \sum_{k=2}^{\infty}(F_{k-1} + F_{k-2})x^k$$

$$= x + \sum_{k=2}^{\infty} F_{k-1}x^k + x^2 \sum_{k=2}^{\infty} F_{k-2}x^k$$

$$= x + x\sum_{k=0}^{\infty} F_k x^k + x^2 \sum_{k=0}^{\infty} F_k x^k$$

$$= x + xs(x) + x^2 s(x).$$

If we put $x = \frac{1}{y}$, then

$$\sum_{k=0}^{\infty} \frac{F_k}{x^k} = \frac{y}{y^2 - y - 1} = \frac{x}{1-x-x^2}.$$

This series converges for $y > 1$.

There are also many series representations related to reciprocals of Fibonacci numbers; for example,

$$\sum_{k=0}^{\infty} \frac{1}{1 + F_{2k+1}} = \frac{\sqrt{5}}{2}.$$

Similarly, the generating function for the Lucas numbers is

$$\sum_{k=0}^{\infty} L_k = \frac{2-x}{1-x-x^2},$$

and we have the various generating functions

$$\sum_{k=0}^{\infty} F_{k+1}x^k = \frac{1}{1 - x - x^2},$$

$$\sum_{k=0}^{\infty} L_{k+1}x^k = \frac{1 + 2x}{1 - x - x^2},$$

$$\sum_{k=0}^{\infty} (-1)^{k+1} F_k x^k = \frac{x}{1 + x - x^2},$$

$$\sum_{k=0}^{\infty} F_{km+r}x^k = \frac{F_r + (-1)^r F_{m-r}x}{1 - L_m x + (-1)^m x^2},$$

$$\sum_{k=0}^{\infty} F_k^2 x^k = \frac{x - x^2}{1 - 2x - 2x^2 + x^3},$$

$$\sum_{k=0}^{\infty} F_k F_{k+1}x^k = \frac{x}{1 - 2x - 2x^2 + x^3},$$

$$\sum_{k=0}^{\infty} L^2 x^k = \frac{4 - 7x - x^2}{1 - 2x - 2x^2 + x^3}.$$

We can use the theory of generating functions to produce explicit forms for the terms of various sequences determined by linear recurrences, as in the following example.

Example 4.1. Consider the sequence

$$\begin{cases} a_0 = 2, \\ a_1 = 3, \\ a_n = 6a_{n-1} - 9a_{n-2} \ (n \geq 2). \end{cases}$$

Then every $a_n = (2 - n) \cdot 3^n$.
   To see this, let

$$g(x) = \sum_{n=0}^{\infty} a_n x^n.$$

Then

$$6xg(x) = \sum_{n=0}^{\infty} 6a_n x^{n+1},$$

and

$$9x^2 g(x) = \sum_{n=0}^{\infty} 9a_n x^{n+2}.$$

It follows that

$$g(x) - 6xg(x) + 9x^2 g(x) = 2 - 9x,$$

or

$$\begin{aligned}
g(x) &= \frac{2 - 9x}{1 - 6x + 9x^2} \\
&= \frac{3}{1 - 3x} - \frac{1}{(1 - 3x)^2} \\
&= 3\sum_{n=0}^{\infty} 3^{n+1} x^n - \sum_{n=0}^{\infty} (n+1)3^n x^n \\
&= \sum_{n=0}^{\infty} (3^{n+1} - (n+1)3^n) x^n \\
&= \sum_{n=0}^{\infty} (2 - n)3^n x^n,
\end{aligned}$$

where we have used the familiar power series expansion

$$\frac{1}{(1-x)^k} = \sum_{n=0}^{\infty} \binom{n+k-1}{k-1} x^n.$$

Using this power series expansion (twice) and the generating function for the Fibonacci sequence, we can also obtain the following identity.

Example 4.2. For every nonnegative integer n,

$$\sum_{k=0}^{n} F_k F_{n-k} = \sum_{2k \le n} k \binom{n-k}{k}.$$

Proof. Set

$$C_n = \sum_{2k \le n} k \binom{n-k}{k}.$$

Then

$$\sum_{n=0}^{\infty} C_n x^n = \sum_{n=0}^{\infty} x^n \sum_{2k \le n} k \binom{n-k}{k}$$

$$= \sum_{k=0}^{\infty} k x^{2k} \sum_{n=0}^{\infty} \binom{n+k}{k} x^n$$

$$= \sum_{k=0}^{\infty} k x^{2k} (1-x)^{-k-1}$$

$$= \frac{x^2}{(1-x)^2} \sum_{k=0}^{\infty} (k+1) x^{2k} (1-x)^{-k}$$

$$= \frac{x^2}{(1-x)^2} \left(1 - \frac{x^2}{(1-x)^2} \right)^{-2}$$

$$= \frac{x^2}{(1-x-x^2)^2}$$

$$= \left(\sum_{n=0}^{\infty} F_n x^n \right) \left(\sum_{j=0}^{\infty} F_j x^j \right).$$

We also obtain in the same way many summation formulas for the Fibonacci numbers and Lucas numbers, and combinations of the two:

$$\sum_{k=0}^{n} \binom{n}{k} F_{k+r} = F_{2n+r},$$

$$\sum_{k=0}^{n} (-1)^{n-k} \binom{n}{k} F_k = (-1)^{n-1} F_n,$$

$$\sum_{k=0}^{n} (-1)^{n-k} F_{2k} = F_n.$$

If the Fibonacci numbers in the three identities above are replaced with Lucas numbers, the results remain valid. A few more examples:

$$\sum_{k=0}^{n} \binom{n}{k} F_k L_{n-k} = 2^n F_{2n},$$

$$\sum_{k=0}^{n} \binom{n}{k} F_k F_{n-k} = \frac{2^n L_n - 2}{5},$$

$$\sum_{k=0}^{n} \binom{n}{k} L_k L_{n-k} = 2^n L_n + 2.$$

□

4.10. Lucas Sequences

Whereas the Fibonacci sequence refers simply to the sequence of Fibonacci numbers, the terminological situation with Lucas sequences is not quite the same. We now introduce general Lucas sequences. Let P and Q be any two nonzero integers, and consider the quadratic polynomial $X^2 - PX + Q$. The discriminant of this polynomial is

$$D = P^2 - 4Q$$

and its roots are given by

$$\alpha, \beta = \frac{P \pm \sqrt{D}}{2},$$

satisfying

$$\begin{cases} \alpha + \beta = P, \\ \alpha\beta = Q, \\ \alpha - \beta = \sqrt{D}. \end{cases}$$

Note that if $D \neq 0$, then necessarily $D \equiv 0$ or $1 \pmod 4$. For $n \geq 0$, we define the sequences

$$U_n(P,Q) = \frac{\alpha^n - \beta^n}{\alpha - \beta},$$

$$V_n(P,Q) = \alpha^n + \beta^n.$$

In particular, always $U_0(P,Q) = 0$, $U_1(P,Q) = 1$, $V_0(P,Q) = 2$, $V_1(P,Q) = P$. The sequence $U(P,Q) = (U_n(P,Q))_{n\geq0}$ is called a *Lucas sequence of the first kind* with parameters P and Q, the sequence $V(P,Q) = (V_n(P,Q))_{n\geq0}$ a *Lucas sequence of the second kind* with parameters P and Q; for fixed P, Q, the two sequences $U(P,Q)$ and $V(P,Q)$ are called *complementary Lucas sequences*.

When $P = 1$, $Q = 1$, then $U(P,Q)$, $V(P,Q)$ are the familiar sequences of Fibonacci numbers and Lucas numbers respectively. In other words, the Fibonacci numbers, and Lucas numbers form a complementary pair of Lucas sequences. When $P = k$ and $Q = -1$, the terms of the sequence $U(k,-1)$ are called k-Fibonacci numbers. In particular, the 2-Fibonacci numbers are also called Lucas–Pell numbers. The first few Lucas–Pell numbers and the first few terms of the complementary sequence are as follows:

$$U(2,-1): \quad 0,1,2,5,12,29,70,169,408,\ldots$$

$$V(2,-1): \quad 2,2,6,14,34,82,198,478,1154,\ldots.$$

When $P = 3$, $Q = 2$, we get the two sequences $U_n(3,2) = 2^n - 1$, $V_n(3,2) = 2^n + 1$, associated respectively with the names and diligent research of Mersenne and Fermat.

If p is an odd prime number, and $\alpha \geq 1$ is any positive integer, then it is known that for all P, Q

$$U_{p^\alpha}(P,Q) \equiv D^{\alpha(p-1)/2} \pmod{p};$$

as a special case, $U_p(P,Q) \equiv (D \mid p) \pmod{p}$. Also, $V_p(P,Q) \equiv P \pmod{p}$.

Conjecture 4.5. *If $U_n = U_n(k,-1)$ is the sequence of k-Fibonacci numbers and p is any prime number, then the first multiple of p among the U_n satisfies $p \equiv (p \mid d) \pmod{n}$, where d is the squarefree part of $k^2 + 4$; that is, d is squarefree and $k^2 + 4 = da^2$ for some integer a. In particular, if $k = 1$, this is a Fibonacci number, and if $k = 2$ this is a Lucas–Pell number with $d = 2$.*

We now use the properties of Lucas sequences to derive the Lucas–Lehmer primality test introduced in Section 1.10. First we need a lemma, which is proved in Section VII (*Mersenne Numbers*) of [8]. Throughout the following lemma and proof, set $P = 2$, $Q = -2$, and let $(U_n)_{n\geq 0}$ and $(V_n)_{n\geq 0}$ be the complementary Lucas sequences $U_n = U_n(P, Q)$, $V_n = V_n(P, Q)$.

Lemma 4.2. *The Mersenne number $M_q = 2^q - 1$ is prime if and only if M_q divides $V_{(M_q+1)/2}$.*

Proof of the Lucas-Lehmer Primality Test. Note that $S_0 = 4 = \frac{V_2}{2}$. Suppose inductively that $S_{k-1} = \frac{V_{2^k}}{2^{2^k-1}}$. Then

$$S_k = S_{k-1}^2 - 2 = \frac{V_{2^k}^2}{2^{2^k}} - 2 = \frac{V_{2^{k+1}} + 2^{2^k+1}}{2^{2^k}} - 2$$

$$= \frac{V_{2^{k+1}}}{2^{2^k}}.$$

Then it follows by Lemma 4.2 that M_n is prime if and only if M_n divides

$$V_{(M_n+1)/2} = V_{2^{n-1}} = 2^{2^{n-2}} S_{n-2},$$

that is, if and only if M_n divides S_{n-2}. $\square$

4.11.   Pisano Period

We now introduce the Pisano period, first observed by Joseph Louis Lagrange (1736–1813) in 1774, and named in honor of Fibonacci, Leonardo Pisano. Let n be any positive integer. Then the nth *Pisano period* $\pi(n)$ is the smallest positive integer such that

$$F_{k+\pi(n)} \equiv F_k \pmod{n}$$

for every integer $k \geq 1$.

It is easy to check by direct computation that $\pi(1) = 1$, $\pi(2) = 3$, $\pi(3) = 8$, $\pi(4) = 6$, $\pi(5) = 20$, $\pi(6) = 24$, $\pi(7) = 16$, $\pi(8) = 12$, $\pi(9) = 24$, $\pi(10) = 60$. For example, the Fibonacci numbers modulo 2 are $(0, 1, 1)$; the Fibonacci numbers modulo 8 are $(0, 1, 1, 2, 3, 5, 0, 5, 5, 2, 7, 1)$.

For $n > 2$, $\pi(n)$ is always even; this follows from Cassini's identity and the observation that necessarily $F_{\pi(n)} \equiv 0 \pmod{n}$, $F_{\pi(n)-1} \equiv F_{\pi(n)+1} \equiv 1 \pmod{n}$. It is also easy to show that $\pi(n) \leq n^2 - 1$ in general. In 1992, Peter Fryed used the theory of finite fields to show that $\pi(n) \le 6n$, with equality for infinitely many n; the smallest such n is $n = 10$, since $\pi(10) = 60$. Also, if p is prime, then it is not difficult to prove that $\pi(p) \le 2p + 2$, where equality is only possible if $(p \mid 5) = 1$. The explicit determination of $\pi(n)$ for arbitrary n is still an open question. It is the case however, that if m divides n, then $\pi(m)$ divides $\pi(n)$.

Even for large n, $\pi(n)$ can sometimes be quite small. For example, $\pi(199) = 22$, $\pi(521) = 26$, $\pi(1364) = 30$, $\pi(3571) = 34$, $\pi(9349) = 38$. In fact, for any positive integer k, $\pi(L_{2k+1}) = 4k+2$. This follows from the identities $F_{-2k-1} = F_{2k+1}$, $L_{2k+1} = F_{2k} + F_{2k+2}$; therefore $F_{-2k} \equiv -F_{2k} \equiv F_{2k+2} \pmod{L_{2k+1}}$, so the Pisano period must divide $4k+2$; but since $F_j < L_{2k+1}$ for all $0 \le j \le 2k+1$, necessarily $\pi(L_{2k+1}) = 4k + 2$. Similarly, every $\pi(F_{2k}) = 4k$, $\pi(F_{2k+1}) = 8k + 4$.

We can define in the same way an analogue $\pi_L(n)$ to the Pisano period for the Lucas numbers; namely, $\pi_L(n)$ is the smallest positive integer such that

$$L_{k+\pi_L(n)} \equiv L_k \pmod{n}$$

for all integers $k \ge 1$. Then it is easy to verify by inspection that $\pi_L(1) = 1$, $\pi_L(2) = 3$, $\pi_L(3) = 8$, $\pi_L(4) = 6$, $\pi_L(5) = 4$, $\pi_L(6) = 24$, $\pi_L(7) = 16$, $\pi_L(8) = 12$, $\pi_L(9) = 24$, $\pi_L(10) = 12$, $\pi_L(11) = 10$, $\pi_L(12) = 24$.

For $n \ge 3$, it is known already that any Fibonacci prime F_n is $F_n \equiv 1 \pmod 4$; for the Lucas numbers, there are primes L_n with both $L_n \equiv 1 \pmod 4$ $(29, 521, \dots)$ and $L_n \equiv 3 \pmod 4$ $(3, 7, 11, 47, 199, \dots)$. We have the following conjecture about the prime divisors of Lucas and Fibonacci numbers.

Conjecture 4.6. *If n is odd, then none of the prime factors of F_n are congruent to 3 modulo 4; if n is even, then at least one of the prime factors of L_n is congruent to 3 modulo 4.*

The first part of this conjecture means that if $n \ge 5$ is an odd number, then F_n has at least one prime factor congruent to 1 modulo 4. For the second part, it is enough to consider only $n \equiv 0 \pmod 6$, since the other cases are trivial. In particular, since the period of the Lucas numbers modulo 4 is $\pi_L(4) = 6$, with cycle $([L_0 = 2], 1, 3, 0, 3, 3, 2, \dots)$, therefore when $n \equiv 2$ or $4 \pmod 6$, $L_n \equiv 3 \pmod 4$ and therefore must have a prime factor $p \equiv 3$

(mod 4). We also have as corollaries to Conjecture 4.7 that F_{2n} has prime factors congruent to 3 modulo 4 and F_{4n} has prime factors congruent to both 1 and 3 modulo 4, for all odd $n \geq 5$.

Although property (3.9) does not hold for the Lucas numbers, we nevertheless have the following result: if L_n is prime, then n must either be zero, prime, or a power of two; to date the only known prime numbers of the form L_{2^k} are $k = 1, 2, 3,$ or 4.

4.12. $\pi(n)$ and $\pi_L(n)$

Suppose $p \neq 5$ is prime, and set

$$\Pi(p) = \begin{cases} 2p + 1 & \text{if } (p \mid 5) = -1, \\ p - 1 & \text{if } (p \mid 5) = 1. \end{cases}$$

Then

$$\pi(p) \text{ divides } \Pi(p). \tag{4.16}$$

Proof of (4.16). Suppose first that $(p \mid 5) = 1$. By (4.4),

$$F_{p-1} \equiv 0 \equiv F_0 \pmod{p},$$
$$F_p \equiv 1 \equiv F_1 \pmod{p}.$$

Therefore $\pi(p)$ divides $p - 1$. Suppose next that $(p \mid 5) = -1$. Then

$$F_{2p+2} = F_{p+1}L_{p+1} \equiv 0 \pmod{p}.$$

Set $n = p + 2$ in the equation $F_{2n-1} = F_n^2 + F_{n-1}^2$. This gives

$$F_{2p+3} = F_{p+2}^2 + F_{p+1}^2$$
$$\equiv F_{p+2}^2$$
$$= (F_{p+1} + F_p)^2$$
$$\equiv F_p^2$$
$$\equiv 1 \pmod{p}.$$

Since $F_0 = 0$, $F_1 = 1$. We conclude that $\pi(p)$ divides $2p + 2$. For the second case, we use the congruences $F_{p+1} \equiv 0 \pmod{p}$, $F_{p+2} \equiv F_p \equiv -1 \pmod{p}$.

Table 4.1 lists the Pisano periods of the first 144 Fibonacci numbers. By inspection, we have $\pi(5) = 20$, and for primes $p \neq 5$, $\pi(p) \leq 2p + 2$. The smallest prime p with $(p \mid 5) = 1$ such that $\pi(p) < \Pi(p)$ is $p = 29$, with $\pi(p = 29) = 14 = \frac{p-1}{2}$; when $(p \mid 5) = -1$, the smallest prime p with $\pi(p) < \Pi(p)$ is $p = 47$, with $\pi(p = 47) = 32 = \frac{2p+2}{3}$.

Table 4.1. Pisano periods.

$\pi(n)$	1	2	3	4	5	6	7	8	9	10	11	12
0+	1	3	8	6	20	24	16	12	24	60	10	24
12+	28	48	40	24	36	24	18	60	16	30	48	24
24+	100	84	72	48	14	120	30	48	40	36	80	24
36+	76	18	56	60	40	48	88	30	120	48	32	24
48+	112	300	72	84	108	72	20	48	72	42	58	120
60+	60	30	48	96	140	120	136	36	48	240	70	24
72+	148	228	200	18	80	168	78	120	216	120	168	48
84+	180	264	56	60	44	120	112	48	120	96	180	48
96+	196	336	120	300	50	72	208	84	80	108	72	72
108+	108	60	152	48	76	72	240	42	168	174	144	120
120+	110	60	40	30	500	48	256	192	88	420	130	120
132+	144	408	360	36	276	48	46	240	32	210	140	24

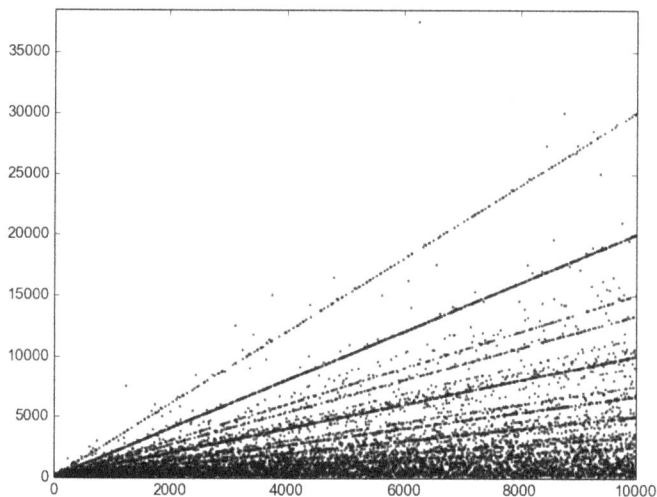

Figure 4.8. Pisano period of the first 10000 Fibonacci numbers

More generally, we have the following conjecture, the first case of which can be proven:

$$\pi_L(n) = \begin{cases} \pi(n) & \text{if 5 does not divide } n, \\ \frac{\pi(n)}{5} & \text{if } n \equiv 0 \pmod 5. \end{cases}$$

In particular $\pi_L(p) = \pi(p)$ for all primes $p \neq 5$. Indeed, suppose n is not a multiple of 5. Then from $L_m = F_{m-1} + F_{m+1}$, it follows that

$$L_{m+\pi(n)} = F_{m+\pi(n)-1} + F_{m+\pi(n)+1}$$
$$\equiv F_{m-1} + F_{m+1}$$
$$= L_m \pmod n$$

for all m. It follows that $\pi_L(n)$ divides $\pi(n)$. Moreover, since $5F_m = L_{m-1} + L_{m+1}$, therefore

$$5F_{m+\pi_L(n)} = L_{m+\pi_L(n)-1} + L_{m+\pi_L(n)+1}$$
$$\equiv L_{m-1} + L_{m+1}$$
$$= 5F_m \pmod n,$$

and since 5 does not divide n we can cancel to obtain $F_{m+\pi_L(n)} \equiv F_m \pmod n$, so also $\pi(n)$ divides $\pi_L(n)$, or $\pi_L(n) = \pi(n)$. □

4.13. Prime Divisors of Lucas Numbers

It is already known that in a single period of the Fibonacci numbers modulo n, the number of zeros can be either 1, 2 or 4, and in a single period of the Lucas numbers modulo n, the number of zeros can be either 0, 1, 2, or 4; in particular, zero may not occur among the Lucas numbers modulo n. For example, no Lucas number is divisible by 5. This shows again that the divisibility property (4.4) does not hold for the Lucas numbers.

In fact, the period of the Lucas numbers modulo 5 is $\pi_L(5) = 4$, with cycle $(2, 1, 3, 4)$; zero also does not occur among the Lucas numbers modulo 13, 17, or 37, with periods 28, 36, and 76, respectively. It is also known in general that if $F_p > 3$ is a Fibonacci prime, then there are no zeros among the Lucas numbers modulo F_p; 17 and 37 are not, however, Fibonacci primes. On the other hand, $L_9 = 76 \equiv 0$ (mod 19), $L_{12} = 322 \equiv 0$ (mod 23), $L_{15} = 1364 \equiv 0$ (mod 31), $L_{10} = 123 \equiv 0$ (mod 41), $L_{22} = 39603 \equiv 0$ (mod 43), $L_{29} \equiv 0$ (mod 59), $L_{34} \equiv 0$ (mod 67), $L_{35} \equiv 0$ (mod 71), $L_{39} \equiv 0$ (mod 79), $L_{42} \equiv 0$ (mod 83).

Therefore, the prime numbers smaller than 100 can be separated into the following different cases:

(1) 2 and 3 are both Fibonacci primes and Lucas primes,
(2) 5, 13, and 89 are Fibonacci primes, and therefore do not divide any Lucas numbers,
(3) 7, 11, 29, 47 are Lucas primes,
(4) 17, 37, 53, 61, 73, and 97 do not divide any Lucas numbers,
(5) 19, 23, 31, 41, 43, 59, 67, 71, 79, and 83 have multiples among the Lucas numbers.

By the periodicity of the Lucas numbers modulo n, if some prime divides any Lucas number, then it divides infinitely many. The question is: what are the prime factors of Lucas numbers? According to our previous analysis, we make the following conjecture.

Conjecture 4.7. *Any prime divisor p of a Lucas number must be $p = 2$, $p \equiv 3$ (mod ()4), or $p \equiv 1$ or 9 (mod 20) where $p = x^2 + 5y^2$ and 4 does not divide xy.*

In 1985, Jeffrey Lagarias used techniques from algebraic number theory to determine that the density of prime factors of Lucas numbers among all prime numbers is $\frac{2}{3}$ (*Pacific J. Math.*, 118.2, 449–461).

Finally, we consider again the Fibonacci primes. We have seen already that except for $F_4 = 3$, if F_n is prime then also n must be prime. It is not known whether or not there are infinitely many Fibonacci primes; to date 33 have been confirmed:

3, 4, 5, 7, 11, 13, 17, 23, 29, 43, 47, 83, 131, 137, 359, 431, 433,
449, 509, 569, 571, 2971, 4723, 5387, 9311, 9677, 14431, 25561,
30757, 35999, 37511, 50833, 81839.

Among the first ten prime numbers, eight are indices of Fibonacci primes; only F_2 and F_{19} are not prime. But after that the Fibonacci primes become more and more rare. There are 1229 prime numbers smaller than 10000, but only 26 of these are indices of Fibonacci primes. The largest Fibonacci prime known today is F_{81839}, with 17103 digits. This was discovered in 2001 by David Broadhurst and Bouk de Water. There are also some Fibonacci numbers suspected to be prime (*probable primes*), corresponding to

$$n = 104911, 130021, 148091, 201107, 397379, 433781,$$
$$590041, 593689, 604711, 931517, 1049897, 1285607,$$
$$1636007, 1803059, 1968721, \text{ and } 2904353.$$

The largest of these is $F_{2904353}$, with 606974 digits, identified as a probable prime by Henri Lifchitz in 2014.

Prime numbers are more common among the Lucas number than among the Fibonacci numbers, but it is still also unknown whether or not there are infinitely many Lucas primes.

Nick MacKinnon proved in 2002 that the only Fibonacci primes that are also one of a twin prime pair are 3, 5, and 13 (*Amer. Math. Monthly*, 109 (2002), p. 78).

Let $\{a_i | 1 \leq i \leq \pi(n)$ be the elements in one cycle of the Pisano period modulo n, it is easy to show that

$$\sum_{i=1}^{\pi(n)} a_i \equiv \sum_{i=1}^{\pi(n)} (-1)^i a_i \equiv \sum_{i=1}^{\pi(n)} a_i^2 \equiv 0 (\text{mod } n)$$

$$\sum_{i=1}^{\pi(n)} a_i^3 \equiv 0 (\text{mod } n), \quad if (n, 10) = 1$$

Moreover, let p be a prime, m be the least positive integer such that p divides F_m. Then

$$\sum_{i=1}^{\pi(p)} a_i = p^2$$

if and only if $m = p + 1$, i.e., $p = 3, 7, 23, 43, 67, 83, \ldots \ldots$

Chapter 5

Perfect Numbers and
Fibonacci Primes

> The true laws cannot be
> linear...
>
> Albert Einstein

5.1. Square Sum Perfect Numbers

In the first chapter, we introduced the perfect numbers and reviewed their history; we also derived a necessary and sufficient condition for an even number to be perfect, namely the Euclid–Euler theorem. This theorem shows that the even perfect numbers are in a one-to-one correspondence with the Mersenne prime numbers, defined in the 17th century. The problem of perfect numbers and Mersenne primes has also become a compelling challenge in the field of computer science. In particular, whether or not there are infinitely many such numbers is an immortal riddle, the oldest and perhaps the most difficult problem in the history of mathematics. In subsequent sections, we also encountered various generalizations of perfect numbers, including the k-multiply perfect numbers, the k-hyperperfect numbers, and so on. Unfortunately, in spite of the efforts of a great many mathematicians, each of these generalizations has given rise only to scattered results without any simple necessary and sufficient conditions.

179

In the spring of 2012, the author of this book proposed after careful consideration and research the *square sum perfect number* problem, concerning numbers satisfying

$$\sum_{\substack{1 \leq d < n \\ d|n}} d^2 = 3n. \tag{5.1}$$

After less than a week, two of my graduate students and I obtained a beautiful result, a necessary and sufficient condition for such numbers involving the Fibonacci sequence of the 13th century. Subsequently, along with three more graduate students, we extended the problem in a more general direction. In this way, we connect the perfect number problem of the early chapters of this book with the Fibonacci sequence, Lucas numbers, and Lucas sequences introduced in the latter two chapters (Fig. 5.1).

We have first the following theorem (Tianxin Cai, Deyi Chen, Yong Zhang, Perfect numbers and Fibonacci primes (I), *Int. J. Number Theory*, **11** (2015) 159–169).

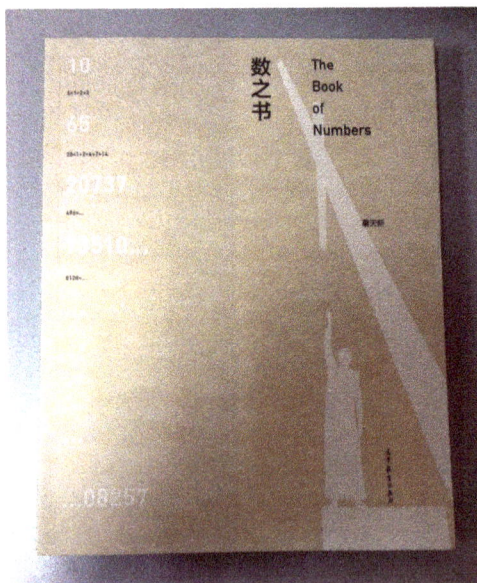

Figure 5.1.   Cover of the Chinese edition of *The Book of Numbers*, and printed on it the first four square sum perfect numbers

Theorem 5.1. *The only integers n satisfying* (5.1) *are all numbers of the form $n = F_{2k-1}F_{2k+1}$ ($k \geq 1$) where both F_{2k-1} and F_{2k+1} are Fibonacci primes.*

If we call the classical perfect numbers M-perfect numbers in light of their affinity with the Mersenne primes, then we can call the numbers satisfying (5.1) F-perfect on account of their connection with Fibonacci primes. The Japanese mathematician Kohji Matsumoto of Nagoya University has also proposed the designations Yin and Yang perfect numbers at the 7th China–Japan Number Theory Conference in Fukuoka in 2013; note that M and F in English also stand for male (yang) and female (yin), respectively.

The largest Fibonacci prime known to date is F_{81839}, the largest Fibonacci probable prime is $F_{1968721}$ (with 411439 digits). Among these there are only five pairs F_{2k-1}, F_{2k+1} of *twin Fibonacci primes*, which generate five F-perfect numbers; these are $n = F_3 F_5$, $F_5 F_7$, $F_{11}F_{13}$, $F_{431}F_{433}$, and $F_{569}F_{571}$ (note that by Corollary 3.1, the indices of any twin Fibonacci primes must be twin primes). Their decimal representations are

10,

65,

20737,

73510803816922669761033626642123533261948011970405233919814585711917444519057612261963528801744523093107269516305744106136707871525711296518385628509088429445930772087319647420825,7,

35232209573904449595952790620404802458842537915400184965695897596126849742246390276402878432136154463286879043721897517251836590479716000271118557285532827829382383900100646042179787559935516043180579182691829284567616114036685771167376,01.

Among them, 10 is the only even F-perfect number; this follows immediately from the fact that 2 is the only even prime number. The fourth and fifth F-perfect numbers have 180 digits and 238 digits, respectively. The next possible F-perfect number has at least 822878 digits; considering that the 51st Mersenne prime, discovered by the GIMPS project, has 24862048 digits, it should be possible to search for larger F-perfect numbers by powerful computer searches. At present however, we do not know even if there is a sixth F-perfect number, nor can we rule out the possibility that there are infinitely many. The situation is just like the situation of the Fermat primes. We can say, however, that if there are infinitely many F-perfect numbers, then the twin prime conjecture must hold.

We introduce next the more general case: if a and b are any positive integers, we consider solutions to the equation

$$\sum_{\substack{1 \le d < n \\ d \mid n}} d^a = bn. \tag{5.2}$$

We have the following theorem.

Theorem 5.2. *If $a = 2$ and $b \ne 3$, or if $a \ge 3$, $b \ge 1$, then (5.2) has at most finitely many solutions; in particular, if $a = 2$ and $b = 1$ or 2, then (5.2) has no solutions.*

In other words, apart from the M-perfect numbers and F-perfect numbers, there appear to be no other interesting perfect numbers along these lines.

5.2.   Some Lemmas

To prove Theorems 5.1 and 5.2, we need the following lemmas.

Lemma 5.1. *If $d > 0$ is odd, then the equation $x^2 - dy^2 = -4$ has solutions in integers if and only if the equation $u^2 - dv^2 = -1$ has solutions in integers.*

Proof.   If $d > 0$ is odd and $x^2 - dy^2 = -4$ for some integers x and y, then x and y must be both even or both odd. Therefore, we can set

$$\begin{cases} u = \dfrac{x(x^2 + 3)}{2}, \\ v = \dfrac{(x^2 + 1)y}{2}. \end{cases}$$

This gives

$$u^2 - dv^2 = \left(\frac{x(x^2 + 3)}{2} \right)^2 - d \left(\frac{(x^2 + 1)y}{2} \right)^2 = -1,$$

as required. The converse is obvious. $\qquad\square$

It is well known that if an integer N is not square, then the continued fraction representation of $\sqrt{N}$ is periodic. We let $l(\sqrt{N})$ denote the period of this continued fraction.

Lemma 5.2. *If $N > 0$ is not square, then the equation $x^2 - Ny^2 = -1$ has solutions in integers if and only if $l(\sqrt{N})$ is odd.*

For the proof, see Kaplan and Williams article (*J. Number Theory*, **23** (1986) 169–182).

Lemma 5.3. *All solutions of the equation $x^2 + y^2 + 1 = 3xy$ with $1 \leq x < y$ are given by*

$$\begin{cases} x = F_{2k-1}, \\ y = F_{2k+1}, \end{cases} \tag{5.3}$$

with $k \geq 1$.

Proof. We prove first that x and y given by (5.3) satisfy $x^2 + y^2 + 1 = 3xy$. By Cassini's identity,

$$F_{2k}^2 - F_{2k+1}F_{2k-1} = -1,$$

so

$$(F_{2k+1} - F_{2k-1})^2 - F_{2k+1}F_{2k-1} = -1,$$

and it follows that

$$1 + F_{2k-1}^2 + F_{2k+1}^2 = 3F_{2k-1}F_{2k+1}.$$

We show next that these are the only solutions. It suffices to show that if $1 + x^2 + y^2 = 3xy$ with $1 \leq x < y$, then $x = F_{2k-1}$ for some $k \geq 1$. Note that this equation is equivalent to

$$5x^2 - 4 = (3x - 2y)^2. \tag{5.4}$$

Then by a result due to Ira Gessel (Fibonacci is a square, *The Fibonacci Quarterly* **10**(4), 1972: 417–419), x is a Fibonacci number. If $x = F_{2k-1}$ for some $k \geq 1$, there is nothing to prove. Suppose $x = F_{2k}$ for some $k \geq 1$. Then equation (5.4) is $5F_{2k}^2 - 4 = (3F_{2k} - 2y)^2$;

then by another result from the same paper by Gessel, necessarily $5F_{2k}^2 + 4 = L_{2k}^2$. We have

$$8 = (5F_{2k}^2 + 4) - (5F_{2k}^2 - 4) = L_{2k}^2 - (3F_{2k} - 2y)^2, \qquad (5.5)$$

which forces $|3F_{2k} - 2y| = 1$, $L_{2k} = 3$. Therefore

$$\begin{cases} x = F_{2k} = \sqrt{\frac{L_{2k}^2 - 4}{5}} = 1 = F_1 = F_2, \\ \\ y = 2 = F_3. \end{cases}$$

We conclude that we always have $x = F_{2k-1}$. $\qquad\square$

Lemma 5.4. *For every positive integer $k \neq 3$, the equation $1 + x^2 + y^2 = kxy$ has no solutions in integers.*

Proof. Since $kxy = 1 + x^2 + y^2 > 2xy$, we can assume $k \geq 4$. If $1 + x^2 + y^2 = kxy$ has solutions in integers, then the discriminant of this quadratic polynomial in x must be a square, so there must exist some integer z such that

$$k^2 y^2 - 4(y^2 + 1) = (k^2 - 4)y^2 - 4 = z^2,$$

or

$$z^2 - (k^2 - 4)y^2 = -4.$$

We show next that k must be odd. Suppose otherwise that k is even; then x and y cannot both be even. If only one of x or y is odd, then reducing $1 + x^2 + y^2 = kxy$ modulo 4 gives $2 \equiv 0 \pmod{4}$, absurd. Similarly, if x and y are both odd, reducing $1 + x^2 + y^2 = kxy$ modulo 2 gives $1 \equiv 0 \pmod 2$.

We assume therefore that k is odd. Evidently, $k^2 - 4$ is not square for odd integers $k \geq 4$, so from Lemmas 5.1 and 5.2, we see that $z^2 - (k^2 - 4)y^2 = -4$ has solutions if and only if $u^2 - (k^2 - 4)v^2 = -1$ has solutions in integers if and only if $l(\sqrt{k^2 - 4})$ is odd. But (see Rosen [9, Exercise 11, on 503])

$$\sqrt{k^2 - 4} = [k - 1; \overline{1, (k-3)/2, 2, (k-3)/2, 1, 2k - 2}].$$

In other words, $l(\sqrt{k^2 - 4}) = 6$; therefore there are no solutions. $\quad\square$

5.3. Proof of Theorems

We now return to Theorems 5.1 and 5.2.

Proof of Theorem 5.2. Let $n = p_1^{\alpha_1} \cdots p_k^{\alpha_k}$ be the prime factorization of n with $p_1 < \cdots < p_k$ and every $\alpha_j \geq 1$; we consider first the case $a = 2$. If $k = 1$, then (5.2) is

$$1 + \sum_{j=1}^{\alpha_1 - 1} p_1^{2j} = bp_1^{\alpha_1},$$

obviously impossible. Suppose $k > 1$. Then by the inequality of arithematic and geometric means, we obtain

$$bn = \sum_{\substack{1 \leq d < n \\ d \mid n}} d^2$$

$$\geq \sum_{j=1}^{k} \frac{n^2}{p_j^2}$$

$$\geq k \left(\prod_{j=1}^{k} \frac{n^2}{p_j^2} \right)^{1/k}$$

$$= n \left(k \prod_{j=1}^{k} p_j^{\alpha_j - \frac{2}{k}} \right).$$

Note that if $k \geq 3$, then every $\alpha_j - \frac{2}{k} \geq \frac{\alpha_j}{3}$; we see that

$$b \geq k \prod_{j=1}^{k} p_j^{\alpha_j/3} \geq 3n^{1/3}. \tag{5.6}$$

This gives an upper bound on the value of n for fixed b, and it follows that there can be at most finitely many solutions. Finally, if $k = 2$, then (5.6) shows that

$$b \geq 2p_1^{\alpha_1 - 1} p_2^{\alpha_2 - 1}, \tag{5.7}$$

from which it follows that α_1, α_2 are bounded above. If $\alpha_2 > 1$, it follows from (5.7) that p_2 is bounded above, so also n is bounded above; on the other hand, if $\alpha_2 = 1$, $\alpha_1 > 1$, then instead p_1 is bounded above. But since

$$\left(\sum_{j=0}^{\alpha_1} p_1^{2j} \right) (1 + p_2^2) - p_1^{2\alpha_1} p_2^2 = b p^{\alpha_1} p_2$$

by (5.2), therefore p_2 divides

$$\sum_{j=0}^{\alpha_1} p_1^{2j}$$

and so also p_2 is bounded above. Therefore, again n is bounded above. Suppose finally that $\alpha_1 = \alpha_2 = 1$. Then

$$1 + p_1^2 + p_2^2 = b p_1 p_2. \tag{5.8}$$

But for $b \neq 3$, it follows from Lemma 5.4 that (5.8) has no solutions in integers.

Summarizing the preceding results, when $a = 2$, $b \neq 3$, then (5.2) has at most finitely many of solutions; in particular, when $b = 1$ or 2, it follows easily from (5.6)–(5.8) that (5.2) has no solutions. We consider finally the case $a \geq 3$; it is easy to see that necessarily $k \neq 1$. When $k \geq 2$, we argue as above that

$$b \geq k \prod_{j=1}^{k} p_j^{(a-1)\alpha_j - \frac{a}{k}} \geq k \prod_{j=1}^{k} p_j^{\alpha_j/2} \geq 2n^{1/2}.$$

This again puts an upper bound on n for fixed b, completing the proof of Theorem 5.2. $\qquad \square$

Proof of Theorem 5.1. Let $n = p_1^{\alpha_1} \cdots p_k^{\alpha_k}$ be the prime factorization of n with $p_1 < \cdots < p_k$ and every $\alpha_j \geq 1$. We show that if n is an F-perfect number, then $k = 2$ and $\alpha_1 = \alpha_2 = 1$. The case $k = 1$ is trivial: in this case, (5.1) reads as

$$\sum_{j=0}^{\alpha_1-1} p_1^{2j} = 3 p_1^{\alpha_1},$$

obviously impossible. If $k \geq 3$, then (5.6) implies that $n = 1$, also impossible. We are left with the case $k = 2$. Suppose that $\alpha_1 + \alpha_2 \geq 3$; noting that

$$\alpha_1^2 - \alpha_1 + 2\alpha_1\alpha_2 \geq \alpha_1(\alpha_1 + \alpha_2),$$
$$\alpha_2^2 - \alpha_2 + 2\alpha_1\alpha_2 \geq \alpha_2(\alpha_1 + \alpha_2),$$

then from the inequality of arithmetic and geometric means, we have

$$3n > \left(\sum_{j=0}^{\alpha_1-1} p_1^{2j}\right) p_2^{2\alpha_2} + \left(\sum_{j=0}^{\alpha_2-1} p_2^{2j}\right) p_1^{2\alpha_1}$$

$$> (\alpha_1 + \alpha_2) \left(\prod_{j=0}^{\alpha_1-1} p_1^{2j} p_2^{2\alpha_2} \cdot \prod_{j=0}^{\alpha_2-1} p_2^{2j} p_1^{2\alpha_1}\right)^{1/(\alpha_1+\alpha_2)}$$

$$\geq 3p_1^{(\alpha_1^2-\alpha_1+2\alpha_1\alpha_2)/(\alpha_1+\alpha_2)} \cdot p_2^{(\alpha_2^2-\alpha_2+2\alpha_1\alpha_2)/(\alpha_1+\alpha_2)}$$

$$\geq 3p_1^{\alpha_1} p_2^{\alpha_2}$$

$$= 3n,$$

which is a contradiction. We conclude that $\alpha_1 = \alpha_2 = 1$, and Theorem 5.1 follows from Lemma 5.3. □

Remark. For $a = 1$, $b > 1$, we have not yet been able to determine whether or not (5.2) has only finitely many solutions.

5.4.   A New Conjecture Concerning Perfect Numbers

Following upon the proofs of Theorems 5.1 and 5.2, we went on to consider similar equations involving cubic analogues to perfect numbers; we were able to prove a theorem, and propose a new conjecture.

Theorem 5.3. *The only positive integer $n = pq$, with p, q distinct primes, such that n divides*

$$\sigma_3(n) = \sum_{d|n} d^3$$

is $n = 6$; more generally, if $n = 2^\alpha p$ for some odd prime p and $\alpha \geq 1$ and n divides $\sigma_3(n)$, then n is a perfect number. Conversely, if $n \neq 28$ is an even perfect number, then n divides $\sigma_3(n)$.

Conjecture 5.1. *If $n = p^\alpha q^\beta$ (p, q distinct primes, α, $\beta \geq 1$) has exactly two prime factors, then n divides $\sigma_3(n)$ if and only if $n \neq 28$ is an even perfect number.*

For the proof of Theorem 5.3, we need a few more lemmas.

Lemma 5.5. *The only primes $p < q$ such that $p|q + 1$ and $q|p + 1$ are $p = 2$, $q = 3$.*

Proof. If p and q satisfy the conditions of the lemma, then it is easy to see that there exists some positive integer k such that $1 + p + q = kpq$. When $k = 1$, this is $1 + p + q = pq$, equivalently $(p-1)(q-1) = 2$; the only solution is $p = 2$, $q = 3$. If $k \geq 2$, then $kpq \geq 2pq > p + q + 1$, so there are no other solutions. $\qquad\square$

Lemma 5.6. *If x and y are positive integers with $y \geq 2$ such that $\frac{x^2 - x + 1}{xy - 1}$ is also a positive integer, then $\frac{x^2 - x + 1}{xy - 1} = 1$.*

Proof. If $x = 1$, $y = 2$, we have $\frac{x^2 - x + 1}{xy - 1} = 1$. Suppose that $\frac{x^2 - x + 1}{xy - 1} = n$ for some $n \geq 2$. Then $x^2 - (1 + ny)x + n + 1 = 0$, so the discriminant must be a square; that is, there must be some integer z such that

$$(1 + ny)^2 - 4(n + 1) = z^2,$$

or

$$\begin{cases} 1 + ny + z = a, \\ 1 + ny - z = b, \end{cases}$$

where $ab = 4(n + 1)$. Therefore $2ny = a + b - 2$. Rearranging,

$$y = \frac{a + b - 2}{2ny} = \frac{a + b - 2}{2(\frac{ab}{4} - 1)} = \frac{2a + 2b - 4}{ab - 4}.$$

We can assume that $a \geq b > 0$. If $b \geq 5$, then

$$y = \frac{2a + 2b - 4}{ab - 4} \leq \frac{2a + 2b - 4}{5a - 4} < 1.$$

Therefore, we need only to consider the cases $b = 1$, 2, 3, or 4:

- if $b = 1$, then $a = 4(n + 1) \geq 12$, so $2 < y = \frac{2a - 2}{a - 4} < 3$;

- if $b = 2$, then $a = 2(n+1) \geq 6$, so $1 < y = \frac{a}{a-2} < 2$;
- if $b = 3$, then $a = \frac{4}{3}(n+1) \geq 4$; if $a \geq 6$, then $y = \frac{2a+2}{3a-4} \leq 1$; if $a = 4$ or 5, then $1 < y = \frac{2a+2}{3a-4} < 2$;
- if $b = 4$, then $a = n+1 \geq 3$; if $a = 3$, we have $y = \frac{a+2}{2a-2} = \frac{5}{4}$; if $a \geq 4$, we have $y = \frac{a+2}{2a-2} \leq 1$.

This completes the proof of Lemma 5.6 □

Lemma 5.7. *The only positive integers $x \geq y$ satisfying $x \mid y^2 - y + 1$ and $y \mid x^2 - x + 1$ are $x = y = 1$.*

Proof. Evidently, if $x = y$, then necessarily both $x = y = 1$. Suppose then that $x > y$. Then we have

$$\begin{cases} y_2 - y + 1 = xt_1, \\ x^2 - x + 1 = yt_2 \end{cases}$$

for some positive integers t_1, t_2. If $t_2 = y$, we see that y divides 1, which forces $x = y = 1$; if $t_2 > y$, then we get

$$y^2 - y + 1 = xt_2 \geq (y+1)^2 = y^2 + 2y + 1,$$

impossible. So we assume $y > t_2$. Then

$$\begin{aligned} t_2^2 y t_1 &= t_2^2(x^2 - x + 1) \\ &= (xt_2)^2 - t_2(xt_2) + t_2^2 \\ &= (y^2 - y + 1)^2 - t_2(y^2 - y + 1) + t_2^2 \\ &= (y^2 - y)^2 + 2(y^2 - y) - t_2(y^2 - y) + t_2^2 - t_2 + 1. \end{aligned}$$

It follows that y divides $t_2^2 - t_2 + 1$, say $t_2^2 - t_2 + 1 = yt_3$. We have

$$\begin{cases} t_2^2 - t_2 + 1 = yt_3, \\ y^2 - y + 1 = t_2 x, \end{cases}$$

where $y > t_2$. Iterating this process, we get a chain of integers

$$x > y > t_2 > t_3 > \cdots > t_k = 1,$$

where the integers $t_2, \ldots t_k$ satisfy $t_{j+1} \mid t_j^2 - t_j + 1$ and $t_j \mid t + j + 1^2 - t_{j+1} + 1$. Since $t_k = 1$, this forces $x = y = t_2 = \cdots = t_k = 1$, contradiction which is a This completes the proof of Lemma 5.7.

□

Proof of Theorem 5.3. For $n = pq$ $(p < q)$, $\sigma_3(n) = 1 + p^3 + q^3 + p^3q^3$. If we assume that n divides $\sigma_3(n)$, this implies that pq divides $1 + p^3 + q^3$, or p divides $q^3 + 1$ and q divides $p^3 + 1$. Factoring $x^3 + 1 = (x+1)(x^2 - x + 1)$, this is equivalent to

$$\begin{cases} p|q+1, \\ q|p+1, \end{cases} \quad \text{or} \quad \begin{cases} p|q+1, \\ q|p^2 - p + 1, \end{cases} \quad \text{or} \quad \begin{cases} p|q^2 - q + 1, \\ q|p+1, \end{cases}$$

$$\text{or} \quad \begin{cases} p|q^2 - q + 1, \\ q|p^2 - p + 1. \end{cases}$$

By Lemma 5.5, the first system has only the solution $p = 2$, $q = 3$, which gives $n = 6$, and by Lemma 5.7 the last system has no solutions. Suppose

$$\begin{cases} p|q+1, \\ q|p^2 - p + 1. \end{cases}$$

Then pq divides $(q+1)(p^2 - p + 1)$, or pq divides $p^2 - p + q + 1$, say $p^2 - p + q + 1 = kpq$, or

$$q = \frac{p^2 - p + 1}{kp - 1} \geq 2.$$

With $k = 1$, this is $q = \frac{p^2 - p + 1}{p+1} = p + \frac{1}{p-1}$ with unique solution $p = 2$, $q = 3$. If $k \geq 2$, then Lemma 5.6 shows that there are no additional solutions. Similarly, the remaining condition

$$\begin{cases} p|q^2 - q + 1 \\ q|p+1 \end{cases}$$

has no solutions. This proves the first claim in the theorem.

We next show that if $n \neq 28$ is an even perfect number, then n divides $\sigma_3(n)$. We verify by direct calculation that 6 divides $\sigma_3(6) = 36$ and 28 does not divides $\sigma_3(28) = 3160$. Suppose $n = 2^{p-1}(2^p - 1)$ where $p \geq 5$ and $2^p - 1$ are both prime, and note that both

$$2^{3p} - 1 = (2^p - 1)(2^{2p} + 2^p + 1)$$

and

$$2^{3p-1} - 1 = 7 \sum_{j=0}^{p-1} 2^{3j}.$$

Therefore 7 divides $2^{2p} + 2^p + 1$. Next we calculate

$$\sigma_3(n) = \sigma_3(2^{p-1}(2^p - 1))$$

$$= \left(\sum_{j=0}^{p-1} 2^{3j} \right) (1 + (2^p - 1)^3)$$

$$= (2^p - 1) \cdot \frac{2^{2p} + 2^p + 1}{7} \cdot 2^p((2^p - 1)^2 - (2^p - 1) + 1)$$

$$= 2n \cdot \frac{2^{2p} + 2^p + 1}{7} \cdot ((2^p - 1)^2 - (2^p - 1) + 1).$$

Therefore n divides $\sigma_3(n)$.

Finally, we show that if $n = 2^{\alpha-1}p$ where $\alpha \geq 2$ and p is an odd prime, and n divides $\sigma_3(n)$, then α must be prime and $p = 2^\alpha - 1$. Note first that

$$\sigma_3(2^{\alpha-1}p) = \sigma_3(2^{\alpha-1})\sigma_3(p)$$

$$= \left(\sum_{j=0}^{\alpha-1} 2^{3j} \right) (1 + p^3)$$

$$= \left(\sum_{j=0}^{\alpha-1} 2^{3j} \right) (1 + p)(1 - p + p^2)$$

$$\equiv 0 \pmod{2^{\alpha-1}p}$$

by hypothesis. It follows that

$$1 + p \equiv 0 \pmod{2^{\alpha-1}}$$

and

$$\sum_{j=0}^{\alpha-1} 2^{3j} \equiv 0 \pmod{p}.$$

So we can find integers k_1 and k_2 such that $p = k_1 2^{\alpha-1} - 1$ and $\frac{2^{3\alpha}-1}{7} = k_2 p$, or

$$2^{3\alpha-1} = (2^\alpha - 1)(2^{2\alpha} + 2^\alpha + 1) = 7k_2(k_1 2^{\alpha-1} - 1). \tag{5.9}$$

If $k_1 = 1$, then $p = 2^{\alpha-1} - 1$; noting that $2^\alpha - 1$ and $2^{\alpha-1} - 1$ are relatively prime, it follows from (5.9) that $2^{2\alpha} + 2^\alpha + 1 \equiv 0$ $(\bmod \ 2^{\alpha-1} - 1)$. Therefore

$$0 \equiv (2^{\alpha-1} - 1)(2^{\alpha+1} + 6) + 7 \equiv 7 \pmod{2^{\alpha-1} - 1}.$$

This forces $\alpha = 4$, $n = 2^{\alpha-1}p = 2^3(2^3 - 1) = 56$. But we can check by hand that 56 does not divide $\sigma_3(56) = 23654$. This rules out the case $k_1 = 1$.

Suppose next that $k_1 \geq 3$. Then by (5.9)

$$2^{2\alpha} + 2^\alpha + 1 \equiv 0 \pmod{k_1 2^{\alpha-1} - 1},$$

so

$$2^{2\alpha} + 2^\alpha + 1 = k_3(k_1 2^{\alpha-1} - 1) \tag{5.10}$$

for some integer k_3. In particular, $2^{\alpha-1}$ divides $k_3 + 1$, say $k_3 = k_4 2^{\alpha-1} - 1$. Substituting in (5.10), we get

$$2^{2\alpha} + 2^\alpha + 1 = \frac{2 + k_1 + k_4}{k_1 k_4 - 4} \geq 2,$$

which implies

$$(k_1 - 1)(k_4 - 1) + k_1 k_4 \leq 11. \tag{5.11}$$

Since $k_1 \geq 3$, we conclude that $k_4 = 1$ or 2.

If $k_4 = 1$, then (5.11) implies $3 \leq k_1 \leq 11$. Noting that

$$2^{\alpha-1} = \frac{2 + k_1 + k_4}{k_1 k_4 - 4} = \frac{k_1 + 3}{k_1 - 4}.$$

This gives either $k_1 = 5$, $\alpha = 4$, $p = k_1 2^{\alpha-1} - 1 = 39$ or $k_1 = 11$, $\alpha = 2$, $p = k_1 2^{\alpha-1} - 1 = 21$. Since neither of 21 or 39 is prime, this is impossible.

If $k_4 = 2$, then by (5.11) $k_1 \leq 4$, or $k_1 = 3$ or 4. In this case

$$2^{\alpha-1} = \frac{2 + k_1 + k_4}{k_1 k_4 - 4} = \frac{k_1 + 4}{2k_1 - 4}.$$

This forces $k_1 = 4$, $\alpha = 2$, $n = 14$. But we check again by hand that 14 does not divide $\sigma_3(14) = 352$; this case too is impossible.

We conclude that $k_1 = 2$, and therefore $p = k_1 2^{\alpha-1} - 1 = 2^\alpha - 1$. In other words, α is prime and n is a perfect number.

In 2018, Xing-Wang Jiang proved (On the even perfect numbers, *Colloq. Math.* **154**(1) 131–136) that Conjecture 5.1 holds when $p = 2$, q is an odd prime; in other words, if $n = 2^\alpha q^\beta$ with q an odd prime and α, $\beta \geq 1$, then n divides $\sigma_3(n)$ if and only if $n \neq 28$ is an even perfect number. In 2019, the graduate student Hao Zhong further proved as part of his doctoral dissertation that if $n = pq^\beta$ where p, q are distinct odd primes, $\beta \geq 1$, and either $q = 3$ or $q \equiv 2$ (mod 3), then n divides $\sigma_3(n)$ if and only if $n \neq 28$ is an even perfect number.

Finally, in 2020, Hung Viet Chu of University of Illinois proved (arXiv:2001.08633v1) the following result: suppose $k > 2$ is prime and $2^k - 1$ is a Mersenne prime, and let $n = 2^{\alpha-1}p$ with $\alpha > 1$ and $p < 3 \cdot 2^{\alpha-1} - 1$ an odd prime; then n divides

$$\sigma_k(n) = \sum_{d|n} d^k$$

if and only if $n \neq 2^{k-1}(2^k - 1)$ is an even perfect number. Moreover, if $n = 2^\alpha p^\beta$ for some integers α, $\beta \geq 1$, then n divides $\sigma_5(n)$ if and only if $n \neq 496$ is an even perfect number. In 2021, Hung Viet Chu further proved (*What's special about the perfect number 6?* Amer. *Math. Monthly* **128**(1) (2021) 87) that if n is an even perfect number, then n divides $\sigma_k(n)$ for every odd $k \geq 1$ if and only if $n = 6$. Xiaoyu Wang has since discovered and proved in addition to this that if n is an even perfect number, then n does not divide $\sigma_k(n)$ for any even $k > 1$.

5.5. Affine Square Sum Perfect Numbers

In the spring of 2013, the author proposed a more general variant on square sum perfect numbers, satisfying the equation

$$\sum_{\substack{1\le d<n \\ d\mid n}} d^2 = An + B. \tag{5.12}$$

When $B = 0$, we have the situation considered in Section 5.1. For $B \neq 0$, we (Tianxin Cai, Liuquan Wang, Yong Zhang, Perfect numbers and Fibonacci primes (II), *Integers*, **19** (A21) (2019) (1–10) obtained the following theorem.

Theorem 5.4.

(1) *If $A = 0$, $B = 1$, then all solutions of (5.12) are given by $n = p$ with p prime.*
(2) *If $A = B = 1$, then all solutions of (5.12) are given by $n = p^2$ with p prime.*
(3) *If $(A, B) \neq (0, 1)$ or $(1, 1)$, then except for finitely many computable solutions in the range $n \le (|A| + |B|)^3$, all solutions of (5.12) are of the form $n = pq$ with $p < q$ primes satisfying the equation*

$$p^2 + q^2 + (1 - B) = Apq. \tag{5.13}$$

For special pairs (A, B), equation (5.13) always has solutions. In particular, if m is any positive integer, then we have the following theorem concerning the Fibonacci and Lucas numbers F_{2m} and L_{2m}.

Theorem 5.5. *Except for finitely many computable solutions in the range*

$$n \le (L_{2m} + F_{2m}^2 - 1)^3,$$

all solutions of the equation

$$\sum_{\substack{1\le d<n \\ d\mid n}} d^2 = L_{2m}n - (F_{2m}^2 - 1) \tag{5.14}$$

are of the form

$$n = F_{2k+1} F_{2k+2m+1}$$

with $k \geq 0$ *and both* F_{2k+1} *and* $F_{2k+2m+1}$ *Fibonacci primes, or the form*

$$n = F_{2k+1} F_{2m-2k-1}$$

with $0 \leq k \neq \frac{m+1}{2} < m$ *and both* F_{2k+1} *and* $F_{2m-2k-1}$ *Fibonacci primes.*

In particular, if $m = 1$ then (5.14) is identical with (5.1), and Theorem 5.5 with Theorem 5.1.

For $1 \leq m \leq 5$. we can use Mathematica to calculate that there are no exceptional solutions to (5.14) other than those given by Fibonacci primes. Known solutions for $1 \leq m \leq 3$ are given in Table 5.1.

When $m = 6$, (5.14) is

$$\sum_{\substack{1 \leq <n \\ d|n}} d^2 = 322n - 20735$$

with exceptional solution $n = 1755 = 3^3 \times 5 \times 13$. We now prove Theorem 5.4.

Proof of Theorem 5.4. The first claim (1) is completely obvious; we assume hereafter that $(A, B) \neq (0, 1)$. Suppose an integer $n > (|A| + |B|)^3$ of (5.12) can be written $n = abc$ for some positive integers $1 < a < b < c$. Then by the inequality of arithmetic and geometric

Table 5.1. Solutions to equation (5.14) for $1 \leq m \leq 3$.

m	$L_{2m} n - (F_{2m}^2 - 1)$	n
1	$3n$	$F_3 F_5,\ F_5 F_7,\ F_{11} F_{13},\ F_{431} F_{433},\ F_{569} F_{571}$
2	$7n - 8$	$F_3 F_7,\ F_7 F_{11},\ F_{13} F_{17},\ F_{43} F_{47}$
3	$18n - 63$	$F_5 F_{11},\ F_7 F_{13},\ F_{11} F_{17},\ F_{17} F_{23},\ F_{23} F_{29},\ F_{131} F_{137}$

means, we have

$$\sum_{\substack{1 \le d < n \\ d \mid n}} d^2 \ge a^2 b^2 + b^2 c^2 + c^2 a^2$$

$$\ge 3(a^4 b^4 c^4)^{1/3}$$

$$= 3n^{4/3}$$

$$> n(|A| + |B|)$$

$$> An + B,$$

so n is not a solution of (5.12); in other words, any solution $n > (|A| + |B|)^3$ of (5.12) has fewer than three distinct nontrivial factors, and in particular at most two distinct prime factors. Let $\omega(n)$ denote the number of distinct prime factors of n. We consider the cases $\omega(n) = 1$ and $\omega(n) = 2$.

First if $\omega(n) = 1$, then $n = p^\alpha$ for some p prime, $\alpha \ge 1$. If $\alpha \ge 6$, we can write $n = p \cdot p^2 \cdot p^{\alpha-3}$ where p, p^2 and $p^{\alpha-3}$ are all distinct integers. Therefore we can consider only $\alpha \le 5$. When $\alpha = 1$, then (5.12) reads as $Ap + B = 1$, with at most one solution $p = \frac{1-B}{A}$. When $\alpha = 2$, (5.12) gives $p^2(1 - A) = B - 1$; with $A = B = 1$, there are infinitely many solutions $n = p^2$ with p any prime number. Otherwise if $A = 1$ and $B \ne 1$, then there are no solutions, and if $A \ne 1$, there is at most one solution $n = p^2 = \frac{B-1}{1-A}$.

If $3 \le \alpha \le 5$, then from (5.12) we see that

$$p(p + p^3 + \cdots + p^{2\alpha-3} - Ap^{\alpha-1}) = B - 1.$$

Note that

$$p + p^3 + \cdots + p^{2\alpha-3} - Ap^{\alpha-1} \equiv p \pmod{p^2},$$

so $p + p^3 + \cdots + p^{2\alpha-3} - Ap^{\alpha-1} \ne 0$ and p^2 divides $B - 1$. For $A = 0$, $B = 2$ or 3, it is east to see that there are no solutions. For every other case we must have

$$n = p^\alpha \le p^5 \le (1 + |B|)^{5/2} \le (|A| + |B|)^3,$$

contradicting the hypothesis.

Next, we consider $\omega(n) = 2$, or $n = p^\alpha q^\beta$. Since n can have at most three distinct nontrivial factors, it is easy to see that we must have both α, $\beta \leq 2$. The case $\alpha = \beta = 2$ is also ruled out for the same reason. If $\alpha = 1$, $\beta = 2$, then

$$\sum_{\substack{1 \leq d < n \\ d \mid n}} d^2 = 1 + p^2 + q^2 + p^2 q^2 + q^4$$

$$= A p q^2 + B$$

$$\leq (|A| + |B|) p q^2.$$

It follows that $p < |A| + |B|$, $q^2 < (|A| + |B|)p$, so

$$n = p q^2 < (|A| + |B|)^2 p < (|A| + |B|)^3,$$

contrary to the hypothesis. By symmetry, there are also no solutions for $\alpha = 2$, $\beta = 1$. Finally, if $\alpha = \beta = 1$, then we obtain exactly the solutions described in claim (3). This completes the proof. $\square$

In the next two lemmas, we collect all the results we need for the proof of Theorem 5.5.

Lemma 5.8. *If $0 \leq m \leq n$ are nonnegative integers, then*

(1) $5F_n^2 + 4(-1)^n = L_n^2$,
(2) $L_m L_n + 5 F_m F_n = 2 L_{m+n}$,
(3) $F_n L_m = F_{n+m} + (-1)^m F_{n-m}$,
(4) $L_n F_m = F_{n+m} - (-1)^m F_{n-m}$,
(5) $5 F_n F_m = L_{n+m} - (-1)^m L_{n-m}$.

Lemma 5.9. *All solutions to the equations $x^2 - 5y^2 = -4$ and $x^2 - 5y^2 = 4$ have the form $(x, y) = (L_{2n+1}, F_{2n+1})$ and $(x, y) = (L_{2n}, F_{2n})$, respectively, for some $n \geq 0$.*

Proof of Theorem 5.5. Note first that (5.13) is equivalent to the equation

$$(2p - Aq)^2 - (A^2 - 4)q = 4(B - 1); \tag{5.15}$$

with $A = L_{2m}$, $B = -F_{2m}^2 + 1$, this is

$$(2p - L_{2m}q)^2 - (L_{2m}^2 - 4)q = -4F_{2m}^2.$$

By Lemma 5.8 (1), this becomes

$$(2p - L_{2m}q)^2 - 5F_{2m}^2 q^2 = -4F_{2m}^2,$$

which implies that F_{2m} divides $2p - L_{2m}q$, say $2p - L_{2m}q = uF_{2m}$. We have $q^2 - 5u^2 = -4$, so by Lemma 5.9, $(u, q) = (\pm L_{2k+1}, F_{2k+1})$ for some nonnegative integer k.

If $(u, q) = (L_{2k+1}, F_{2k+1})$, then $p = \frac{1}{2}(L_{2m}F_{2k+1} + L_{2k+1}F_{2m})$, so by (3) and (4) of Lemma 5.8, $p = F_{2k+2m+1}$, and therefore $n = F_{2k+1}F_{2k+2m+1}$, where both F_{2k+1} and $F_{2k+2m+1}$ are prime.

If on the other hand $(u, q) = (-L_{2k+1}, F_{2k+1})$, then $p = \frac{1}{2}(L_{2m}F_{2k+1} - L_{2k+1}F_{2m})$. If $2k + 1 > 2m$, then by (3) and (4) of Lemma 5.8, we get $p = F_{2k-2m+1}$, $n = F_{2k+1}F_{2k-2m+1}$; if $2k + 1 < 2m$, then we get instead that $p = F_{2m-2k-1}$, $n = F_{2k+1}F_{2m-2k-1}$. This completes the proof. □

5.6.   Square Sum Perfect Numbers and the Twin Prime Conjecture

In the previous section, we considered affine square sum perfect numbers, which are determined by a combination of Fibonacci numbers and Lucas numbers. We have analogous results using only the Lucas numbers. We present below two theorems, which can both be proved using Lemma 5.8(2) and (5); for details see Tianxin Cai, Liuquan Wang, Yong Zhang, Perfect numbers and Fibonacci primes (II), *Integers*, **19** (A21) (2019) 1–10.

Theorem 5.6. *Except for finitely many computable solutions in the range*

$$n \leq (L_{2m}^2 + L_{2m}^2 - 3)^3,$$

all solutions to the equation

$$\sum_{\substack{1 \leq d < n \\ d|n}} d^2 = L_{2m}n - (L_{2m}^2 - 3) \tag{5.16}$$

are of the form $n = L_{2k-1}L_{2k+2m-1}$, where both L_{2k-1} and $L_{2k+2m-1}$ are prime.

Theorem 5.7. *Except for finitely many computable solutions in the range*

$$n \leq (L_{2m}^2 + L_{2m}^2 - 5)^3,$$

all solutions to the equation

$$\sum_{\substack{1 \leq d < n \\ d \mid n}} d^2 = L_{2m}n - (L_{2m}^2 - 5)$$

are either

(1) $n = L_{2k}L_{2k+2m}$ *with* $k \geq 0$ *and both* L_{2k} *and* L_{2k+2m} *prime, or*
(2) $n = L_{2k}L_{2m-2k}$ *with* $0 \leq k \neq \frac{m}{2} \leq m$ *and both* L_{2k} *and* L_{2m-2k} *prime.*

For $1 \leq m \leq 5$, we can verify with Mathematica that there are no exceptional solutions to (5.16) other than those given by the theorem. Table 5.2 lists known solutions for $1 \leq m \leq 3$. The solutions $n = L_{613}L_{617}$ and $L_{4787}L_{4793}$ have 258 and 2003 digits respectively.

In 1849, the French mathematician Alphonse de Polignac (1826–1863) proposed the following conjecture.

Conjecture 5.2 (de Polignac). *For every positive integer* k, *there exist infinitely many prime numbers* p *and* q *such that* $p - q = 2k$.

When $k = 1$, this is the famous *twin prime conjecture*.

Returning the proof of Theorem 5.4 in the previous section, we obtain also from it the following result.

Theorem 5.8. *Let* A *and* k *be any positive integers, and consider the equation*

$$\sum_{\substack{1 \leq d < n \\ d \mid n}} d^2 = An + (k^2 + 1). \tag{5.17}$$

Table 5.2. Solutions to equation (5.16) for $1 \leq m \leq 3$.

m	$L_{2m}n + (L_{2m}^2 - 3)$	n
1	$3n + 6$	L_5L_7, $L_{11}L_{13}$, $L_{17}L_{19}$
2	$7n + 46$	L_7L_{11}, $L_{13}L_{17}$, $L_{37}L_{41}$, $L_{613}Ls_{617}$
3	$18n + 321$	L_5L_{11}, L_7L_{13}, $L_{11}L_{17}$, $L_{13}L_{19}$, $L_{31}L_{37}$, $L_{41}L_{47}$, $L_{47}L_{53}$, $L_{4787}L_{4793}$

(1) *If $A \neq 2$ or k is odd, then* (5.17) *has only finitely many solutions.*
(2) *If $A = 2$ and k is even, then except for finitely many computable solutions in the range $n < (|A| + k^2 + 1)^3$, all solutions are of the form $n = p(p + k)$ with both p and $p + k$ prime.*

Corollary 5.1. *The equation*

$$\sum_{\substack{1 \leq d < n \\ d \mid n}} d^2 = 2n + 4k^2 + 1$$

has infinitely many solutions for every $k \geq 1$ if and only if de Polignac's conjecture holds.

Considering only the case $k = 1$, we have the following special case.

Corollary 5.2. *The equation*

$$\sum_{\substack{1 \leq d < n \\ d \mid n}} d^2 = 2n + 5$$

has infinitely many solutions if and only if the twin prime conjecture holds.

We also find that the equation

$$\sum_{\substack{1 \leq d < n \\ d \mid n}} 2kd^2 - (2k-1)d = (4k^2 + 1)n + 2$$

has infinitely many solutions for $k \geq 1$ if and only there are infinitely many primes p such that also $2kp + 1$ is prime. When $k = 1$, this condition amounts to the existence of infinitely many Sophie Germain primes. Finally, there are infinitely many primes p such that $2p - 1$ is also prime if and only the if the equation

$$\sum_{\substack{1 \leq d < n \\ d \mid n}} 2kd^2 - (2k-1)d = (4k^2 + 1)n + 4k$$

has infinitely many solutions.

Proof of Theorem 5.8. Suppose $n > (|A| + k^2 + 1)^3$ is a solution of (5.17); then it follows from Theorem 5.4(3) that $n = pq$ for distinct

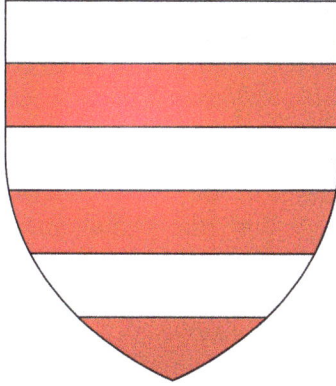

Figure 5.2. Family crest of the de Polignac family

primes p and q satisfying $p^2 + q^2 - k^2 = Apq$. We assume without loss of generality that $p < q$ and note that q divides $(p - k)(p + k)$.

If $p = k$, we get that $q = Ak$ and $n = Ak^2 < (|A| + k^2 + 1)^3$; if $p < k$, then $p + k < 2k$, and since $p < q$, we must have that q divides $p + k$, so $n < 2k^2 < (|A| + k^2 + 1)^3$.

Therefore, we consider only primes $p > k$. We have that q divides $p + k$ and $2q > 2p > p + k$. It follows that $q = p + k$, and therefore

$$A = \frac{p^2 - q^2 - k^2}{pq} = 2,$$

$n = p(p + k)$. If $p \geq 3$, we must have k even.

Conversely, if $A = 2$, and both p and $p + k$ are prime, then it is easy to verify that $n = p(p + k)$ is a solution of (5.17). $\qquad\square$

As an aside, it is worth mentioning that the de Polignac family included many prominent figures in French history; Jules de Polignac became the first Duke of Polignac in the late 18th century, and Alphonse de Polignac's father served as prime minister under Charles X (Fig. 5.2).

5.7. Fermat Primes and GM Numbers

In Section 1.6, we discussed prime numbers of the form $M_p = 2^p - 1$, called the Mersenne primes, which are known to be in bijective correspondence with the even perfect numbers, as we have seen. In 1640, four years before Mersenne began to consider the numbers which bear his name, Fermat put forward a similar equation:

$$F_n = 2^{2^n} + 1.$$

The numbers F_n have since come to be called *Fermat numbers*, and prime numbers of this form are known as *Fermat primes*. Fermat himself verified by hand that the Fermat numbers $F_0 = 3$, $F_1 = 4$, $F_2 = 17$, $F_3 = 257$, and $F_4 = 65536$ are all prime. He conjectured therefore that F_n is prime for all integers $n \geq 0$.

This conjecture went unresolved for nearly a century until Leonard Euler proved that F_6 in 1732, when he was 25 years old and living in St. Petersburg. Specifically, he showed that 641 divides F_5, arguing as follows. Set $a = 2^7$, $b = 5$. Then $a - b^3 = 3$ amd $1 + ab - b^4 = 1 + 3b = 2^4$; so

$$F_5 = (2a)^4 + 1$$
$$= (1 + ab - b^4)a^4 + 1$$
$$= (1 + ab)a^4 + 1 - a^4 b^4$$
$$= (1 + ab)(a^4 + (1 - ab)(1 + a^2 b^2)).$$

We conclude that $1 + ab = 641$ divides F_5. Since then, mathematicians have checked F_n for more than 40 values of n, including all $5 \leq n \leq 32$, and no new Fermat primes have been discovered among them, although nobody has been able to determine explicity the prime factors of F_{20} and F_{24}. The situation is therefore quite different from the situation of the Mersenne numbers, where new Mersenne primes have been discovered in a more or less uninterrupted stream for centuries. It is known however from the identity

$$F_n = 2 + \prod_{k=0}^{n-1} F_k$$

that any two Fermat numbers are relatively prime; this was first observed by Goldbach.

Therefore, there remain still many unsolved mysteries related to the Fermat numbers. Is every F_n with $n > 4$ composite? Are there infinitely many composite Fermat numbers? Infinitely many Fermat primes? This last question was first formulated explicitly by the

German Ferdinand Eisenstein (1823–1852) in 1844. Since the resolution of Fermat's last theorem, the problem of Fermat primes could be said to deserve this designation.

We consider also generalized Fermat numbers of the form $a^{2^n} + b$ with a an even number and b an odd number. It is not difficult to verify that the first six numbers $2^{2^n} + 15$ ($0 \leq n \leq 5$) are prime: these are 17, 19, 31, 271, 65551, 4293967391. Otherwise, there is not much to say: we also do not know whether there are infinitely many composite generalized Fermat numbers or infinitely many prime generalized Fermat numbers.

In 2014, inspired by the perfect numbers, the author introduced the *GM numbers*, which satisfy the equation

$$n = 2^\alpha + \sum_{\substack{1 \leq d < n \\ d \mid n}} d,$$

or

$$\sigma(n) = 2n - 2^\alpha,$$

for some positive integer $\alpha \geq 1$. Note that if n is an odd prime, then

$$\sum_{\substack{1 \leq d < n \\ d \mid n}} d = 1,$$

and we see that the only prime GM numbers are the Fermat primes; therefore we can view the GM numbers as a generalization of the Fermat primes.

The source of the name for these numbers comes from the fact that the time when the author began thinking about this question happened to coincide with the passing away of the Colombian novelist Gabriel Garcia Márquez (1927–2014), whose masterpiece *One Hundred Years of Solitude* calls to mind those difficult mathematical problems that have taken long centuries of effort to overcome (Figs. 5.3 and 5.4). We put forward on the platform Weibo the question whether or not there are any odd composite GM numbers, and follower Alpha discovered two. These are

$$19649 = 7^2 \times 401 = 2^{14} + 3265, \text{ and}$$

$$22075325 = 5^2 \times 883013 = 2^{24} + 5298109.$$

Figure 5.3. The first Spanish edition (1967) of *One Hundred Years of Solitude*

It has since been verified that these are the only composite GM smaller than 2×10^{10}. In total, there are seven known odd GM numbers, including the Fermat primes. As for even GM numbers, they are much more numerous. There are six already among the numbers smaller than 100. These are:

$$10, 14, 22, 38, 44, 92;$$

for example,

$$10 = 2 + 1 + 2 + 5,$$
$$14 = 2^2 + 1 + 2 + 7.$$

If we consider numbers smaller than 10^6 and 10^8, we find 146 and 350 even GM numbers, respectively.

Figure 5.4.   Gabriel Garcia Márquez, in a turban

More generally, if we consider prime numbers of the form $2^\alpha + 2^\beta - 1$ for some integers $\alpha, \beta \geq 1$, the list includes all Fermat primes and all Mersenne primes. There are two questions, or rather conjectures to consider:

(1) there are infinitely many prime numbers of the form $2^\alpha + 2^\beta - 1$ ($\alpha, \beta \geq 1$),
(2) there are infinitely many even GM numbers.

It is easy to see that (2) follows immediately from (1): if $p = 2^\alpha + 2^\beta - 1$ is prime, then both $2^{\alpha-1}p$ and $2^{\beta-1}p$ are even GM numbers.

We have also another related question. Recall by way of binary notation that every odd prime number admits a unique representation as a sum

$$1 + \sum_{j=1}^{k} 2^{n_j}$$

with $1 \leq n_1 < \cdots < n_k$. We ask whether or not there are infinitely many prime numbers that admit a representation of this kind for a given fixed length $k \geq 1$. This is a generalization of Eisenstein's question concerning Fermat primes, which corresponds to the case $k = 1$. If we define the *length* of a prime number

$$p = 1 + \sum_{j=1}^{k} 2^{n_j}$$

to be length$(p) = k$, then the primes of length 1 are 2 and every Fermat prime; the primes of length 2 are 7, 11. 13, 19, 41, $\ldots$; the primes of length 3 are 23, 29, 43, 53, $\ldots$; and so on.

If on the other hand we define the *weight* of the above prime p to be prime number

$$p = 1 + \sum_{j=1}^{k} 2^{n_j}$$

the weight of the above prime p to be

$$\text{weight}(p) = \sum_{j=1}^{k} n_k,$$

we can classify the prime numbers also according to their weight. The weights 1, 2, 3 each correspond to only a single prime number, 3, 5, and 7, respectively; there are two primes each of weights 4 and 5, these are 11 and 17, and 13 and 19, respectively; there are no primes of weight 6; the weights 7 and 8 both correspond to three primes, 23, 37, and 67, and 41, 131, 257, respectively; and so on. We pose the following the question: are there any positive integers other than 6 that are not the weight of any prime number?

In addition to the above questions and speculation, we also have some positive results. For example, if $n > 1$ is odd, then $1+2+2^{2^n}$ is composite; also, for any integer $j > 1$, there is a unique integer $k \geq 1$ such that $j \equiv 2^{k-1} + 1 \pmod{2^k}$, and it follows that $1 + 2^j + 2^{2^n}$ is composite for all $n \geq k$. So there are infinitely many integer pairs m, n such that $1 + 2^{2^m} + 2^{2^n}$ is composite. In fact, if $n > 1$, then $1 + 2 + 2^{2^n}$ is always a multiple of 7; and if $j = 2^{k-1} + 1 + 2^k t$ for some positive integers k, t, then

$$1 + 2^j + 2^{2^n} = 1 + 2(2^{2^{k-1}})^{2t+1} + (2^{2^{k-1}})^{2^{m-k+1}}$$

$$\equiv 1 - 2 + 1 = 0 \pmod{F_{k-1}},$$

where F_{k-1} is the $(k-1)$-th Fermat number.

Passing from base 2 to base 3, we consider prime numbers of the form

$$\frac{3^{2^n} + 1}{2}.$$

When $n = 0$, 1, 2, or 4 we have, respectively, the primes 2, 5, 41, 2152361; $n = 3$ gives a composite number $3281 = 17 \times 193$. In comparison, primes of the form $3^{2^n} + 2$ are somewhat more numerous.

5.8.   The *abcd* Equation

In early 2013, the author casually and capriciously introduced *the abcd equation*, defined below, not expecting that it would prove to have deep connections with the Fibonacci numbers. Research into this equation has since shown to involve rich research methodology, and the difficulty of its solution is inestimable. Even if a solution is found, it will still be worthwhile to consider the number and structure of its solutions. In (Tianxin Cai, Chen Deyi, *On the abcd Problem*), we present a preliminary discussion of these issues.

Definition 5.1. Let n be a positive integer, and a, b, c, d take values among the positive rational numbers. The *abcd* equation is

$$n = (a + b)(c + d), \tag{5.18}$$

where a, b, c, d are required to satisfy $abcd = 1$.

By the inequality of arithmetic and geometric means, $(a + b)(c + d) \geq 2\sqrt{ab} \times 2\sqrt{cd} = 4$, so there are no solutions when $n = 1$, 2, or 3. On the other hand, for $n = 4$ or 5, we have the solutions

$$4 = (1 + 1)(1 + 1),$$

$$5 = (1 + 1)\left(2 + \frac{1}{2}\right).$$

Moreover, any single solution of (5.18) generates an infinite family of solutions of the form $\left(ka, kb, \frac{c}{k}, \frac{d}{k}\right)$,

It is easy to see in general that if (5.18) has solutions among positive rational numbers, then the equation

$$n = x + \frac{1}{x} + y + \frac{1}{y} \tag{5.19}$$

also has solutions in positive rational numbers, and conversely. This is because

$$x + \frac{1}{x} + y + \frac{1}{y} = (x+y)\left(1 + \frac{1}{xy}\right).$$

In particular, when $n = 4$ or 5, the solutions $x = 1$, $y = 1$ and $x = 2$, $y = 2$, respectively, of (5.19) are unique. The first of these claims is obvious, the second requires the theory of elliptic curves to prove.

Theorem 5.9. *If either 8 divides n or $\mathrm{ord}_2(n) = 1$, then equation (5.19) has no solutions; also, if n is odd, or $\mathrm{ord}_2(n) = 2$, and n has at least one prime factor $p \equiv 3 \pmod 4$, then (5.19) has no solutions.*

Proof. Suppose (5.19) admits solutions for given n; then we can write

$$n = \frac{a}{b} + \frac{b}{a} + \frac{c}{d} + \frac{d}{c}$$

for positive integers a, b, c, d satisfying $\gcd(a,b) = \gcd(c,d) = 1$. Multiplying both sides by ab, we get

$$abn - (a^2 + b^2) = \frac{ab}{cd}(c^2 + d^2).$$

Since $\gcd(cd, c^2 + d^2) = 1$, we conclude that cd divides ab, since otherwise the right-hand side cannot be an integer. By the same argument, ab divides cd. We conclude that $ab = cd$.

Therefore, we have

$$abn = a^2 + b^2 + c^2 + d^2 = (a \pm b)^2 + (c \mp d)^2. \tag{5.20}$$

Arguing from parity, symmetry, and the identity $ab = cd$, we can consider only two cases for (a,b,c,d):

(1) Case 1: (odd, odd, odd, odd),
(2) Case 2: (odd, even, odd, even).

Since the square of any even number is 0 modulo 4 and the square of any odd number is 1 modulo 8, it follows that if 8 divides n or

$\mathrm{ord}_2(n) = 1$, then (5.20) has no solutions, and therefore also the *abcd* equation has no solutions.

If n is odd or $\mathrm{ord}_2(n) = 2$, then from the theory of quadratic residues we find that the left-hand side of (5.20) cannot have any prime factors $p \equiv 3 \pmod 4$. For suppose otherwise that some $p \equiv 3 \pmod 4$ divides the left-hand side of (5.20), and note p cannot divide both $c + d$ and $c - d$, since this implies p divides $\gcd(c, d) = 1$. But if p does not divide $c \pm d$, then $\left(-(c + d)^2 \mid p\right) = -1$, where $(\cdot \mid \cdot)$ is the Legendre symbol, which is impossible. This completes the proof.

$\square$

We consider next the equation

$$ n = \left(a + \frac{1}{a}\right)\left(b + \frac{1}{b}\right), \tag{5.21} $$

where a and b are positive integers. Obviously, if this equation has solutions for some n, then also the *abcd* equation has solutions. We find that equation (5.21) yields integers if and only if $\gcd(a, b) = 1$ and

$$ \begin{cases} a \mid b^2 + 1, \\ b \mid a^2 + 1, \end{cases} $$

or

$$ a^2 + b^2 + 1 \equiv 0 \pmod{ab}; $$

that is, if there exists some integer $q \geq 1$ such that

$$ a^2 + b^2 + 1 = qab. $$

From Lemmas 5.3 and 5.4, we know that this equation has solutions if and only if $q = 3$, in which case the only integral solutions of (5.21) are given by $a = F_{2k-1}$, $b = F_{2k+1}$ for some $k \geq 1$, where the F_n are Fibonacci numbers. These remarks are summarized as Theorem 5.10.

Theorem 5.10. *If $n = F_{2k-3}F_{2k+3}$ for some $k \geq 1$, then the abcd equation admits the solution $a = F_{2k-1}$, $b = F_{2k+1}$ in (5.21).*

It follows immediately that there are infinitely many integers n such that the *abcd* equation admits solutions, among them $n = 4, 5, 13, 68, 445, 3029, 20740, \ldots$.

Via the Pisano period introduced in Section 4.11, we also get the following theorem.

Theorem 5.11. *If n is an odd integer satisfying (5.21), then necessarily $n \equiv 5 \pmod{8}$; If n is an even integer satisfying (5.21), then necessarily $n = 4m$ for some $m \equiv 1 \pmod{16}$.*

Proof.   It follows as we have seen from the proof of Theorem 5.10 that integers satisfying (5.21) have the form $n = F_{2k-3}F_{2k+3}$ for some $k \geq 1$. In Section 4.11, we observed that the Pisano period modulo 8 is 12, with residue cycle

$$(1, 1, 2, 3, 5, 0, 5, 5, 2, 7, 1, 0).$$

It follows that the numbers $n = F_{2k-3}F_{2k+3}$ are periodic in k with period 6. When $k = 3$ or 6, n is even; when $k = 1, 2, 4$ or 5, then $n \equiv 5 \pmod{8}$. It follows easily that if n is an even number satisfying (5.21), then $n = F_{6k-3}F_{6k+3}$, and since the Fibonacci numbers satisfy $F_{6k+3} \equiv 2 \pmod{32}$, we get that $n \equiv 4 \pmod{64}$, proving the theorem.   □

Another related equations is

$$n = \left(\frac{a}{b} + \frac{b}{a}\right)\left(\frac{c}{d} + \frac{d}{c}\right), \tag{5.22}$$

where a, b, c, d are positive integers satisfying $\gcd(a, b) = \gcd(c, d) = 1$. It is obvious that equation (5.21) above is a special case of equation (5.22), and that solutions to (5.22) furnish solutions to both (5.18) and (5.19). In fact, the converse also holds: given a solution (5.18), put $x = \frac{a}{b}$, $y = \frac{c}{d}$, and note that following the proof of Theorem 5.9, $ab = cd$. Substituting in (5.19), we get

$$n = (x + y)\left(1 + \frac{1}{xy}\right) = \left(\frac{a}{c} + \frac{c}{a}\right)\left(\frac{c}{b} + \frac{b}{c}\right),$$

a solution of (5.22). Hereafter we refer to any of (5.18), (5.19) or (5.22) as the *abcd* equation. As for (5.22), we have so far obtained only partial results. For example, when $b = 1$, $d = 2$, and $2c$ divides $a^2 + 1$, a divides $c^2 + 4$, we have the solutions

$$1237 = \left(\frac{17}{1} + \frac{1}{17}\right)\left(\frac{145}{2} + \frac{2}{145}\right),$$

$$6925 = \left(\frac{337}{1} + \frac{1}{337}\right)\left(\frac{41}{2} + \frac{2}{41}\right),$$

including the prime 1237. When a, c, d are odd, $b = 1$, and cd divides $a^2 + 1$, a divides $c^2 + d^2$, we have the solutions

$$580 = \left(\frac{157}{1} + \frac{1}{157}\right)\left(\frac{5}{17} + \frac{17}{5}\right),$$

$$1156 = \left(\frac{73}{1} + \frac{1}{73}\right)\left(\frac{13}{205} + \frac{205}{13}\right),$$

$$5252 = \left(\frac{697}{1} + \frac{1}{697}\right)\left(\frac{5}{37} + \frac{37}{5}\right),$$

$$32976266756 = \left(\frac{33169}{1} + \frac{1}{33169}\right)\left(\frac{17}{257} + \frac{257}{17}\right).$$

A more interesting result: we can generate an infinite family of solutions to (5.21) or (5.22), therefore also the *abcd* equation. For example, put $(x_0, y_0) = (1, 1)$, $(41, 137)$, or $(386, 35521)$ and consider the sequences containing (x_0, y_0) such that every adjacent three numbers x, y, z satisfy $xz = y^4 + 1$:

$$\dots, 41761, 17, 2, x_0 = 1, y_0 = 1, 2, 17, 41761, \dots,$$

$$\dots, 20626, x_0 = 41, y_0 = 137, 8592082, \dots,$$

$$\dots, 624977, x_0 = 386, y_0 = 35531, \dots.$$

Then every two adjacent numbers x, y in each sequence generate a solution to (5.22) given by $a = x^2 + y^2$, $b = 1$, $c = x$, $d = y$ corresponding to

$$n = \frac{(c^2 + d^2) + 1}{cd}.$$

Theorem 5.12. *Suppose a solution n of the abcd equation is odd or satisfies $\mathrm{ord}_2(n) = 2$, so n has no prime factors $p \equiv 3 \pmod 4$ by Theorem 5.9; then if $p \equiv 4 \pmod 4$ divides $n \pm 4$, then $\mathrm{ord}_p(n \pm 4)$ is even.*

Proof. From the proof of Theorem 5.9, we see that if (5.19) has solutions, then there are integers positive a, b, c, d such that $\gcd(a, b) = \gcd(c, d) = 1$, $ab = cd$, and

$$nab = (a \pm b)^2 + (c \mp d)^2,$$

and it follows easily from the theory of quadratic residues that the left-hand side of this equation, and in particular both a and b, have no prime factors $p \equiv 3 \pmod 4$. Turning to $n \pm 4$, we also have

$$(n \pm 4)\, ab = (a \pm b)^2 + (c \pm d)^2. \tag{5.23}$$

Suppose first that some prime $p \equiv 3 \pmod 4$ divides $n + 4$; if p does not divide $c + d$, it follows from the theory of quadratic residues that (5.23) cannot hold; therefore we can assume that p divides $c + d$, so also p divides $a + b$; it follows that p^2 divides $n + 4$; by the same argument, if p^k divides $n + 4$ with k odd, we must have that p^{k+1} also divides $n + 4$, so $\mathrm{ord}_p(n + 4)$ is even. The case $p \equiv 3 \pmod 4$ divides $n - 4$ is similarly handled. $\qquad\square$

Corollary 5.3. *If* $n = F_{2k-3}F_{2k+3}$ *for some integer* $k \geq 0$, *then either* n *is odd or* $\mathrm{ord}_2 n = 2$, *and* n *has no prime factors* $p \equiv 3 \pmod 4$; *if some prime* $p \equiv 3 \pmod 4$ *divides* $n \pm 4$, *then* $\mathrm{ord}_p(n \pm 4)$ *is even.*

Corollary 5.4. *If* $n = 4m$ *for some* $m \geq 1$ *admits solutions to the* abcd *equation, then* $m \equiv 1 \pmod 8$.

Proof. From Theorem 5.9, we see that $m \equiv 1 \pmod 4$. If $m \equiv 5 \pmod 8$, say $m = 8k + 5$, then $n + 4 = 8(4k + 3)$, so $n + 4$ has at least one prime factor $p \equiv 3 \pmod 4$ with $\mathrm{ord}_p(n + 4)$ odd, contradicting Theorem 5.12. Therefore $m \equiv 1 \pmod 8$. $\qquad\square$

From Theorems 5.9, 5.10, and 5.12 we conclude that except for $n = 4$, 5, 13, 68, 445, and 580, the integers not positive exceeding 1000 that might admit solutions to the *abcd* equation are $n = 41$, 85, 113. 149, 229, 265, 292, 365, 373, 401, 481, 545, 761, 769, 797, 877, 905, and 932.

Conjecture 5.3. *The positive integers* $n \equiv 1 \pmod 8$ *do not admit solutions to the* abcd *equation.*

Conjecture 5.4. *If $n = 4m$ for some $m \geq 1$ admits solutions to the abcd equation, then $m \equiv 1 \pmod{16}$.*

Remark. If we assume Conjectures 5.3 and 5.3, we reduce the possible positive integers less than 1000 for which the *abcd* equation can have solutions to $n = 85, 149, 229, 365, 373, 797$, and 877.

Question 5.1. Are there infinitely many positive integers for which the *abcd* equation has a solution but equation (5.21) does not?

5.9. Applications of Elliptic Curves

We now consider the number of solutions to (5.19), when solutions exist. If $n = 4$, it is easy to verify via the inequality of arithmetic and geometric means that (5.19) has a unique solution. If $n > 4$, we transform (5.19) into an elliptic curve.

Theorem 5.13. *If $n > 4$, equation (5.19) admits solutions if and only if the elliptic curve*

$$E_n : \; Y^2 = X^3 + (n^2 - 8)X^2 + 16X$$

has rational points with $X < 0$ (see illustration) (Fig.5.5).

Proof. Define $x, y \geq 1$ such that $x + y > 2$ via the variables s, $t > 0$ as

$$\begin{cases} x = \dfrac{s + nt}{2(t + t^2)}, \\[2mm] y = \dfrac{s + nt}{2(1 + t)}. \end{cases} \tag{5.24}$$

Then we have the inverse transformation

$$\begin{cases} s = \dfrac{2y^2 + (2x - n)y}{x}, \\[2mm] t = \dfrac{y}{x}; \end{cases}$$

we conclude that (5.24) determines a bijection. Substituting into (5.19) and simplifying, we get

$$n = \frac{(s + nt)^2 + 4t(1 + t^2)^2}{2t(s + nt)}.$$

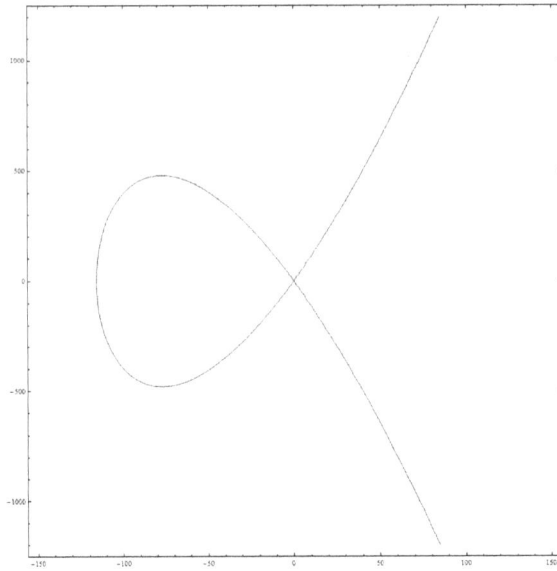

Figure 5.5.   The elliptic curve $E(13)$.

Now set

$$
\begin{cases}
X = -4t, \\
Y = 4s.
\end{cases}
$$

This produces the elliptic curve E_n and completes the proof.   □

Note that we can write x and y above in terms of X and Y as

$$
\begin{cases}
x = \dfrac{2Y - 2nX}{X^2 - 4X}, \\
y = \dfrac{Y - nX}{2(4 - X)}.
\end{cases}
\tag{5.25}
$$

Example 5.1. The equation

$$
5 = x + \frac{1}{x} + y + \frac{1}{y}
$$

has unique positive rational solution $x = y = 2$.

By Theorem 5.13, we can see this by identifying rational points with $X < 0$ of the elliptic curve

$$E_5 : Y^2 = X^3 + 25X^2 + 16X.$$

Using the Magma Computational Algebra System (Magma), we find that the rank of E_5 is zero. By Mordell's famous theorem, the rational points of E_5 form a finitely generated abelian group $E_5(\mathbb{Q})$ satisfying

$$E_5(\mathbb{Q}) \cong E_5(\mathbb{Q})_{\text{tor}} \oplus \mathbb{Q}^r,$$

where $E_5(\mathbb{Q})_{\text{tor}}$ is the torsion part of $E_5(\mathbb{Q})$. We can obtain all the rational points by computation; they are $(-16, 0)$, $(-4, -12)$, $(-4, 12)$, $(-1, 0)$, $(0, 1)$, $(4, -20)$, $(4, 20)$ and the point at infinity. Checking these in (5.25) we see that the only solution is $x = y = 2$, obtained from the point $(-4, 12)$.

For general n, we can obtain several new theorems using the properties of elliptic curves and the theory of torsion points (see [2]).

Theorem 5.14. *If the abcd equation has any positive rational solution for some $n \geq 6$, then it has infinitely many.*

Example 5.2. The equation

$$13 = x + \frac{1}{x} + y + \frac{1}{y} \tag{5.26}$$

has infinitely many positive rational solutions.

To see this we again invoke Theorem 5.13, and study the elliptic curve

$$E_{13} : Y^2 = X^3 + 161X^2 + 16X.$$

when $X < 0$. Magma finds that the rank of this elliptic curve is 1; via Mordell's theorem, we find a generator $P(X, Y) = (-100, 780)$. Substituting in (5.26), we easily obtain the first positive rational solution $(\frac{2}{5}, 10)$; the second and third are given by

$$(x, y) = \left(\frac{924169}{228730}, \frac{1347965}{156818} \right),$$

$$(x, y) = \left(\frac{33896240819350898}{3149745790659725}, \frac{12489591059767450}{8548281631402489} \right).$$

Theorem 5.14 shows that there are infinitely many more.

5.10.   Results with Lucas Sequences

We return in this section to equation (5.12):

$$\sum_{\substack{1 \le d < n \\ d \mid n}} d^2 = An + B,$$

where A and B are any integers. Using the definitions and notation of Section 4.10, we have the following theorem.

Theorem 5.15. *Let P take values in integers. Except for finitely many computable solutions in the range $n \le (|A| + |B|)^3$, all integer solutions of (5.12) have one of the following forms:*

(1) $n = U_{2k-1}(P, -1)U_{2k+1}(P, -1)$ *with $A = P^2 + 2$, $B = -P^2 + 1$, and both $U_{2k-1}(P, -1)$ and $U_{2k+1}(P, -1)$ prime,*

(2) $n = U_{2k}(P, -1)U_{2k+2}(P, -1)$ *with $A = P^2 + 2$, $B = P^2 + 1$, and both $U_{2k}(P, -1)$ and $U_{2k+2}(P, -1)$ prime, or*

(3) $n = U_{k-1}(P, 1)U_{k+1}(P, 1)$ *with $A = P^2 - 2$, $B = P^2 + 1$, and both $U_{k-1}(P, 1)$ and $U_{k+1}(P, 1)$ prime.*

If we put $P = 1$ in (2), we get Theorem 5.1; if we put $P = 2$ in (3), we get Corollary 2 to Theorem 5.8. We also have the following similar result.

Theorem 5.16. *Let P take values in integers. Except for finitely many computable solutions in the range $n \le (|A| + |B|)^3$, all integer solutions of (5.12) have one of the following forms:*

(1) $n = V_{2k}(P, -1)V_{2k+2}(P, -1)$ *with $A = P^2 + 2$, $B = -P^4 - 4P^2 + 1$, where $P^2 + 4$ is squarefree and both $V_{2k}(P, -1)$ and $V_{2k+2}(P, -1)$ are prime,*

(2) $n = V_{2k-1}(P, -1)U_{2k+1}(P, -1)$ *with $A = P^2 + 2$, $B = P^4 + 4P^2 + 1$, where $P^2 + 4$ is squarefree and both $V_{2k-1}(P, -1)$ and $V_{2k+1}(P, -1)$ are prime or*

(3) $n = V_{k-1}(P, 1)V_{k+1}(P, 1)$ *with $A = P^2 + 2$, $B = -P^4 + 4P^2 + 1$, where $P^2 - 4$ is squarefree and both $V_{k-1}(P, 1)$ and $V_{k+1}(P, 1)$ are prime.*

In the same way, we can generalize Theorems 5.6, 5.7 and 5.8 using Lucas sequences.

The proofs of Theorems 5.15 and 5.16 require a few lemmas. Lemmas 5.10 and 5.14 below appears in *On the Lucas property of linear recurrent sequences* (Hao Zhong and Tianxin Cai, *Int. J. Number Theory*, **13** (6), (2017) 1617–1625); Lemma 5.11 in *Enumerable Sets are Diophantine* (*Doklady Akademii Nauk SSSR*, **191** (1970) 279—282); Lemmas 5.12 and 5.13 appear in *Representation of solutions of Pell equations using Lucas sequences* (Jones, J. P., *Acta Academiae Paedagogicae Agriensis, Sectio Mathematicae* **30** (2003) 75–86). Throughout, the numbers A_n are obtained from the recurrence

$$A_n = uA_{n-1} + vA_{n-2}$$

for $n \geq 2$, with A_0, A_1, u, v fixed arbitrary integers.

Lemma 5.10. *For arbitrary nonnegative integers $n \geq r \geq 0$,*

$$A_{n+r}A_{n-r} - A_n^2 = (-v)^{n-r}s^2(r-1,u,v)(vA_0^2 + uA_0A_1 - A_1^2),$$

where

$$s(k,u,v) = \sum_{j=0}^{\lfloor k/2 \rfloor} \binom{k-j}{j} u^{k-2j}v^j.$$

If we fix $u = 1$ in Lemma 5.10, we obtain the following identities: if $v = 1$, then

$$1 + A_{2n}^2 + A_{2n+2}^2$$
$$= (u^2 + 2)A_{2n}A_{2n+2} - u^2(A_0^2 + uA_0A_1 - A_1^2) + 1,$$
$$1 + A_{2n-1}^2 + A_{2n+1}^2$$
$$= (u^2 + 2)A_{2n-1}A_{2n+1} + u^2(A_0^2 + uA_0A_1 - A_1^2) + 1;$$

if $v = -1$, then

$$1 + A_{n-1}^2 + A_{n+1}^2 = (u^2 - 2)A_{n-1}A_{n+1} + u^2(A_0^2 + uA_0A_1 - A_1^2) + 1,$$

From the above identities, it is easy to see that integers of the form $n = pq$ as in Theorems 5.15 and 5.16 are solutions of (5.12).

Lemma 5.11. *All solutions in positive integers of the equation $x^2 - (P^2 + 4)y^2 = 4$ have the form $x = V_{2k}(P, -1)$, $y = U_{2k}(P, -1)$.*

Lemma 5.12. *All solutions in positive integers of the equation* $x^2 - (P^2 + 4)y^2 = -4$ *have the form* $x = V_{2k+1}(P, -1)$, $y = U_{2k+1}(P, -1)$.

Lemma 5.13. *All solutions in positive integers of the equation* $x^2 - (P^2 - 4)y^2 = 4$ *have the form* $x = V_k(P, 1)$, $y = U_k(P, -1)$.

Lemma 5.14. *For any integer* k,

(1) $V_k(P, Q) = U_{k+l}(P, Q) - QU_{k-l}(P, Q)$;
(2) $(P^2 - 4Q)U_k(P, Q) = V_{k+l}(P, Q) - QV_{k-l}(P, Q)$.

Proof of Theorem 5.15. (1) Note that we can transform equation (5.13) in Theorem 5.4 into

$$(2p - Aq)^2 - (A^2 - 4)q^2 = 4(B - 1). \qquad (5.27)$$

Put $(A, B) = (P^2 + 2. - P^2 + 1)$ in (5.27). This gives

$$(2p - (P^2 + 2)q)^2 - P^2(P^2 + 4)q^2 = -4P^2.$$

If we let $r = (2p - (P^2 + 2)q)/P$, we can write this as

$$r^2 - (P^2 + 4)q^2 = -4.$$

Therefore, by Lemma 5.12, we require

$$(r, q) = (\pm V_{2k+1}(P, -1), U_{2k+1}(P, -1))$$

for some integer k. Similarly, from Lemma 5.13, we have $p = U_{2k-1}(P, -1)$ or $U_{2k+3}(P, -1)$. This gives

$$n = U_{2k-1}(P, -1)U_{2k+1}(P, -1),$$

where both $U_{2k-1}(P, -1)$ and $U_{2k+1}(P, -1)$ are prime.

(2) Put instead $(A, B) = (P^2 + 2. P^2 + 1)$ in (5.27). This gives

$$(2p - (P^2 + 2)q)^2 - p^2(P^2 + 4)q^2 = 4P^2.$$

If we let $r = (2p - (P^2 + 2)q)/P$, we can write this as

$$r^2 - (P^2 + 4)q^2 = 4.$$

Therefore, by Lemma 5.11, we require

$$(r, q) = (\pm V_{2k}(P, -1), U_{2k}(P, -1))$$

for some integer k. Using Theorem 5.14 and the recurrence formula for Lucas sequences, we consider separately the two cases:

Case 1: $(r, q) = (V_{2k}(P, -1), U_{2k}(P, -1))$, in which case

$$
\begin{aligned}
2p &= rP + (P^2 + 2)q \\
&= PV_{2k}(P, -1) + (P^2 + 2)U_{2k}(P, -1) \\
&= PU_{2k-1}(P, -1) + PU_{2k+1}(P, -1) + (P^2 + 2)U_{2k}(P, -1) \\
&= P(U_{2k+1}(P, -1) - PU_{2k}(P, -1)) + PU_{2k+1}(P, -1) \\
&\quad + (P^2 + 2)U_{2k}(P, -1) \\
&= 2PU_{2k+1}(P, -1) + 2U_{2k}(P, -1) \\
&= 2U_{2k+2}(P, -1),
\end{aligned}
$$

or $p = U_{2k+2}(P, -1)$;

Case 2: $(r, q) = (-V_{2k}(P, -1), U_{2k}(P, -1))$, in which case

$$
\begin{aligned}
2p &= rP + (P^2 + 2)q \\
&= -PV_{2k}(P, -1) + (P^2 + 2)U_{2k}(P, -1) \\
&= -PU_{2k-1}(P, -1) - PU_{2k+1}(P, -1) + (P^2 + 2)U_{2k}(P, -1) \\
&= -PU_{2k-1}(P, -1) - P(PU_{2k}(P, -1)) + PU_{2k-1}(P, -1)) \\
&\quad + (P^2 + 2)U_{2k}(P, -1) \\
&= 2U_{2k}(P, -1) - 2PU_{2k-1}(P, -1) \\
&= 2U_{2k-2}(P, -1),
\end{aligned}
$$

or $p = U_{2k-2}(P, -1)$.

We conclude that

$$
n = U_{2k}(P, -1)U_{2k+2}(P, -1),
$$

where both $U_{2k}(P, -1)$ and $U_{2k+2}(P, -1)$ are prime.

(3) Finally, put $(A, B) = (P^2 - 2.P^2 + 1)$ in (5.27). This gives

$$(2p - (P^2 - 2)q)^2 - P^2(P^2 - 4)q^2 = 4P^2.$$

If we let $r = (2p - (P^2 - 2)q)/P$, we can write this as

$$r^2 - (P^2 - 4)q^2 = 4.$$

Therefore, by Lemma 5.13, we require

$$(r, q) = (\pm V_k(P, 1), U_k(P, 1))$$

for some integer k. Arguing as above, we conclude from Lemma 5.14(1) that $p = U_{k-2}(P, 1)$ or $p = U_{k+2}(P, 1)$. We conclude that

$$n = U_{k-1}(P, 1)U_{k+1}(P, 1),$$

where both $U_{k-1}(P, 1)$ and $U_{k+1}(P, 1)$ are prime. This completes the proof of Theorem 5.15. $\square$

Proof of Theorem 5.16 (1) put $(A, B) = (P^2 + 2, -P^4 - 4P^2 + 1)$ in (5.27). this gives

$$(2p - (P^2 + 2)q)^2 - P^2(P^2 + 4)q^2 = -4P^2(P^2 + 4).$$

If we let

$$r = \frac{2p - (P^2 + 2)q}{P(P^2 + 4)},$$

we can write this as

$$q^2 - (P^2 + 4)r^2 = 4,$$

where we use the requirement that $P^2 + 4$ is squarefree. Therefore, by Lemma 5.11, we require

$$(r, q) = (\pm U_{2k}(P, -1), V_{2k}(P, -1))$$

for some integer k. Using Lemma 5.13 and the relevant recurrence relations, we consider separately the two cases:

Case 1: $(r, q) = (U_{2k}(P, -1), V_{2k}(P, -1))$, in which case

$$2p = rP(P^2 + 4) + (P^2 + 2)q$$
$$= P(P^2 + 4)U_{2k}(P, -1) + (P^2 + 2)V_{2k}(P, -1)$$
$$= PV_{2k-1}(P, -1) + PV_{2k+1}(P, -1) + (P^2 + 2)V_{2k}(P, -1)$$
$$= P(V_{2k+1}(P, -1) - PV_{2k}(P, -1)) + PV_{2k+1}(P, -1)$$
$$+ (P^2 + 2)V_{2k}(P, -1)$$
$$= 2PV_{2k+1}(P, -1) + 2V_{2k}(P, -1)$$
$$= 2V_{2k+2}(P, -1),$$

or $p = V_{2k+2}(P, -1)$;

Case 2: $(r, q) = (-U_{2k}(P, -1), V_{2k}(P, -1))$, in which case

$$2p = rP(P^2 + 4) + (P^2 + 2)q$$
$$= -P(P^2 + 4)U_{2k}(P, -1) + (P^2 + 2)V_{2k}(P, -1)$$
$$= -PV_{2k-1}(P, -1) - PV_{2k+1}(P, -1) + (P^2 + 2)V_{2k}(P, -1)$$
$$= -PV_{2k-1}(P, -1) - P(PV_{2k}(P, -1)) + V_{2k-1}(P, -1))$$
$$+ (P^2 + 2)V_{2k}(P, -1)$$
$$= 2V_{2k}(P, -1) - 2PV_{2k-1}(P, -1)$$
$$= 2V_{2k-2}(P, -1),$$

or $p = V_{2k-2}(P, -1)$.

We conclude that

$$n = V_{2k}(P, -1)V_{2k+2}(P, -1)$$

where both $V_{2k}(P, -1)$ and $V_{2k+2}(P, -1)$ are prime.

(2) Put instead $(A, B) = (P^2 + 2.P^4 + 4P^2 + 1)$ in (5.27). This gives

$$(2p - (P^2 + 2)q)^2 - P^2(P^2 + 4)q^2 = 4P^2(P^2 + 4).$$

If we let

$$r = \frac{2p - (P^2 + 2)q}{P(P^2 + 4)},$$

we can write this as

$$q^2 - (P^2 + 4)r^2 = -4,$$

where we use the requirement that $P^2 + 4$ is squarefree. Therefore, by Lemma 5.12, we require

$$(r, q) = (\pm U_{2k+1}(P, -1), V_{2k+1}(P, -1))$$

for some integer k. From Lemma 5.14(2), we get $p = V_{2k-1}(P, -1)$ or $V_{2k+3}(P, -1)$. This gives

$$n = V_{2k-1}(P, -1)V_{2k+1}(P, -1)$$

where both $V_{2k-1}(P, -1)$ and $V_{2k+1}(P, -1)$ are prime.

(3) Finally, put $(A, B) = (P^2 - 2, -P^4 + 4P^2 + 1)$ in (5.27). This gives

$$(2p - (P^2 - 2)q)^2 - P^2(P^2 - 4)q^2 = -4P^2(P^2 - 4).$$

If we let

$$r = \frac{2p - (P^2 - 2)q}{P(P^2 - 4)},$$

we can write this as

$$q^2 - (P^2 - 4)r^2 = 4,$$

where we use the requirement that $P^2 - 4$ is squarefree. Therefore, by Lemma 5.13, we require

$$(r, q) = (\pm U_k(P, 1), V_k(P, 1))$$

for some integer k. From Lemma 5.14(2), we get $p = V_{2k-2}(P, 1)$ or $V_{k+2}(P, 1)$. This gives

$$n = V_{k-1}(P, 1)V_{k+1}(P, 1),$$

where both $V_{k-1}(P, 1)$ and $V_{k+1}(P, 1)$ are prime. This completes the proof of Theorem 5.16. □

The proofs of above theorems appear in Perfect Numbers and Fibonacci prime III (Hao Zhong, Tianxin Cai, preprint).

Finally, considering de Polignac's conjecture and its generalization by Dickson (Dickson's conjecture), and, in turn, the generalization of its generalization by Sierpinski and Schinzel, we would like to know if there are equivalent or similar generalizations of Theorems 5.8, 5.15 and 5.16. Or, is there a quadratic polynomial $f \in \mathbb{Z}[x]$ such that the solutions of

$$\sum_{\substack{1 \leq d < n \\ d \mid n}} d^2 = f(n)$$

can be expressed in some interesting way with reference to the prime numbers (excluding perhaps finitely many computable solutions, as above).

The great 20th century physicist Albert Einstein (1879–1955) wrote in his autobiographical notes that "the true laws cannot be linear, nor can they be derived from linearity...." This claim was perhaps the product of excessive exuberance following his discovery of the mass–energy conversion formula of the special theory of relativity, but it has proven valid as a description of the results in this book.

Appendix

Appendix A.1.　The First 100 Fibonacci Numbers and Their Prime Factorizations

n	F_n	Prime Factorization of F_n
1	1	1
2	1	1
3	2	2
4	3	3
5	5	5
6	8	2^3
7	13	13
8	21	$3 \cdot 7$
9	34	$2 \cdot 17$
10	55	$5 \cdot 11$
11	89	89
12	144	$2^4 \cdot 3^2$
13	233	233
14	377	$13 \cdot 29$
15	610	$2 \cdot 5 \cdot 61$
16	987	$3 \cdot 7 \cdot 47$
17	1,597	1597
18	2,584	$2^3 \cdot 17 \cdot 19$
19	4,181	$37 \cdot 113$
20	6,765	$3 \cdot 5 \cdot 11 \cdot 41$

(*Continued*)

Appendix A.1. (*Continued*)

n	F_n	Prime Factorization of F_n
21	10,946	$2 \cdot 13 \cdot 421$
22	17,711	$89 \cdot 199$
23	28,657	28657
24	46,368	$25 \cdot 32 \cdot 7 \cdot 23$
25	75,025	$52 \cdot 3001$
26	121,393	$233 \cdot 521$
27	196,418	$2 \cdot 17 \cdot 53 \cdot 109$
28	317,811	$2 \cdot 13 \cdot 29 \cdot 281$
29	514,229	514229
30	832,040	$2^3 \cdot 5 \cdot 11 \cdot 31 \cdot 61$
31	1,346,269	$577 \cdot 2417$
32	2,178,309	$3 \cdot 7 \cdot 47 \cdot 2207$
33	3,524,578	$2 \cdot 89 \cdot 19801$
34	5,702,887	$1597 \cdot 3571$
35	9,227,465	$5 \cdot 13 \cdot 141961$
36	14,930,352	$2^4 \cdot 3^3 \cdot 17 \cdot 19 \cdot 107$
37	24,157,817	$73 \cdot 149 \cdot 2221$
38	39,088,169	$37 \cdot 113 \cdot 9349$
39	63,245,986	$2 \cdot 233 \cdot 135721$
40	102,334,155	$3 \cdot 5 \cdot 7 \cdot 11 \cdot 41 \cdot 2161$
41	165,580,141	$2789 \cdot 59369$
42	267,914,296	$2^3 \cdot 13 \cdot 29 \cdot 211 \cdot 421$
43	433,494,437	433494437
44	701,408,733	$3 \cdot 43 \cdot 89 \cdot 199 \cdot 307$
45	1,134,903,170	$2 \cdot 5 \cdot 17 \cdot 61109441$
46	1,836,311,903	$139 \cdot 461 \cdot 28657$
47	2,971,215,073	2971215073
48	4,807,526,976	$2^6 \cdot 3^2 \cdot 7 \cdot 23 \cdot 47 \cdot 1103$
49	7,778,742,049	$13 \cdot 97 \cdot 6168709$
50	12,586,269,025	$5^2 \cdot 11 \cdot 101 \cdot 151 \cdot 3001$
51	20,365,011,074	$2 \cdot 1597 \cdot 6376021$
52	32,951,280,099	$3 \cdot 233 \cdot 521 \cdot 90481$
53	53,316,291,173	$953 \cdot 55945741$
54	86,267,571,272	$2^3 171953 \cdot 1095779$
55	139,583,862,445	$5 \cdot 89 \cdot 661 \cdot 474541$
56	225,851,433,717	$3 \cdot 7^2 \cdot 13 \cdot 29 \cdot 281 \cdot 14503$
57	365,435,296,162	$2 \cdot 37 \cdot 113 \cdot 797 \cdot 54833$
58	591,286,729,879	$59 \cdot 19489 \cdot 514229$
59	956,722,026,041	$353 \cdot 2710260697$
60	1,548,008,755,920	$2^4 \cdot 3^2 \cdot 5 \cdot 11 \cdot 31 \cdot 41 \cdot 61 \cdot 2521$

(*Continued*)

Appendix A.1. (*Continued*)

n	F_n	Prime Factorization of F_n
61	2,504,730,781,961	$4513 \cdot 555003497$
62	4,052,739,537,881	$557 \cdot 2417 \cdot 3010349$
63	6,557,470,319,842	$2 \cdot 13 \cdot 17 \cdot 421 \cdot 35239681$
64	10,610,209,857,723	$3 \cdot 7 \cdot 47 \cdot 1087 \cdot 2207 \cdot 4481$
65	17,167,680,177,565	$5 \cdot 233 \cdot 14736206161$
66	27,77,7890,035,288	$2^3 \cdot 89 \cdot 199 \cdot 9901 \cdot 19801$
67	44,915,570,212,853	$269 \cdot 116849 \cdot 1429913$
68	72,723,460,248,141	$3 \cdot 67 \cdot 1597 \cdot 3571 \cdot 63443$
69	117,669,030,460,994	$2 \cdot 137 \cdot 829 \cdot 18077 \cdot 28657$
70	190,392,490,709,135	$511 \cdot 13 \cdot 29 \cdot 71 \cdot 911 \cdot 141961$
71	308,061,521,170,129	$6673 \cdot 46165371073$
72	498,454,011,879,264	$2^5 \cdot 3^3 \cdot 7 \cdot 17 \cdot 19 \cdot 23 \cdot 107 \cdot 103681$
73	806,515,533,049,393	$9375829 \cdot 86020717$
74	1,304,969,544,928,657	$73 \cdot 149 \cdot 2221 \cdot 54018521$
75	2,111,485,077,978,050	$2 \cdot 5^2 \cdot 61 \cdot 3001 \cdot 230686501$
76	3,416,454,622,906,707	$3 \cdot 37113 \cdot 9349 \cdot 29134601$
77	5,527,939,700,884,757	$13 \cdot 89 \cdot 988681 \cdot 4832521$
78	8,944,394,323,791,464	$2^3 \cdot 79 \cdot 233 \cdot 521 \cdot 859 \cdot 135721$
79	14,472,334,024,676,221	$157 \cdot 92180471494753$
80	23,416,728,348,467,685	$3 \cdot 5 \cdot 7 \cdot 11 \cdot 41 \cdot 47 \cdot 1601 \cdot 2161 \cdot 3041$
81	37,889,062,373,143,906	$2 \cdot 17 \cdot 53 \cdot 109 \cdot 2269 \cdot 4373 \cdot 19441$
82	61,305,790,721,611,591	$2789 \cdot 59369 \cdot 370248451$
83	99,194,853,094,755,497	99194853094755497
84	160,500,643,816,367,088	$2^4 \cdot 3^2 \cdot 13 \cdot 29 \cdot 83 \cdot 211 \cdot 281 \cdot 421 \cdot 1427$
85	259,695,496,911,122,585	$5 \cdot 1597 \cdot 9521 \cdot 3415914041$
86	420,196,140,727,489,673	$6709 \cdot 144481 \cdot 433494437$
87	679,891,637,638,612,258	$2 \cdot 173 \cdot 514229 \cdot 3821263937$
88	1,100,087,778,366,101,931	$3 \cdot 7 \cdot 43 \cdot 89 \cdot 199 \cdot 263 \cdot 307 \cdot 881 \cdot 967$
89	1,779,979,416,004,714,189	$1069 \cdot 1665088321800481$
90	2,880,067,194,370,816,120	$2^3 \cdot 5 \cdot 11 \cdot 1719 \cdot 31 \cdot 61181 \cdot 541 \cdot 109441$
91	4,660,046,610,375,530,309	$13^2 \cdot 233 \cdot 741469159607993$
92	7,540,113,804,746,346,429	$3 \cdot 139 \cdot 461 \cdot 4969 \cdot 28657 \cdot 275449$
93	12,200,160,415,121,876,738	$2 \cdot 557 \cdot 2417 \cdot 4531100550901$
94	19,740,274,219,868,223,167	$2971215073 \cdot 6643838879$
95	31,940,434,634,990,099,905	$5 \cdot 37 \cdot 113 \cdot 761 \cdot 9641 \cdot 67735001$
96	51,680,708,854,858,323,072	$2^7 \cdot 3^2 \cdot 7 \cdot 23 \cdot 47 \cdot 769 \cdot 1103 \cdot 2207 \cdot 3167$
97	83,621,143,489,848,422,977	$193 \cdot 389 \cdot 3084989 \cdot 3610402019$
98	135,301,852,344,706,746,049	$13 \cdot 29 \cdot 97 \cdot 6168709 \cdot 599786069$
99	218,922,995,834,555,169,026	$2 \cdot 17 \cdot 89 \cdot 197 \cdot 19801 \cdot 18546805133$
100	354,224,848,179,261,915,075	$3 \cdot 5^2 \cdot 11 \cdot 41101 \cdot 151 \cdot 401 \cdot 3001 \cdot 570601$

Appendix A.2.   The First 100 Lucas Numbers and Their Prime Factorizations

n	L_n	Prime Factorization of L_n
1	1	1
2	3	3
3	4	2^2
4	7	7
5	11	11
6	18	$2 \cdot 3^2$
7	29	29
8	47	47
9	76	$2^2 \cdot 19$
10	123	$3 \cdot 41$
11	199	199
12	322	$2 \cdot 7 \cdot 23$
13	521	521
14	843	$3 \cdot 281$
15	1,364	$22 \cdot 11 \cdot 31$
16	2,207	2207
17	3,571	3571
18	5,778	$2 \cdot 3^3 \cdot 107$
19	9,349	9349
20	15,127	$7 \cdot 2161$
21	24,476	$2^2 \cdot 29 \cdot 211$
22	39,603	$3 \cdot 43 \cdot 307$
23	64,079	$139 \cdot 461$
24	103,682	$2 \cdot 47 \cdot 1103$
25	167,761	11101151
26	271,443	$3 \cdot 90481$
27	439,204	$2^2 \cdot 19 \cdot 5779$
28	710,647	$7^2 \cdot 4503$
29	1,149,851	$59 \cdot 19489$
30	1,860,498	$2 \cdot 3^2 \cdot 41 \cdot 2521$
31	3,010,349	3010349
32	4,870,847	$1087 \cdot 4481$
33	7,881,196	$2^2 \cdot 199 \cdot 9901$
34	12,752,043	$3 \cdot 67 \cdot 63443$
35	20,633,239	$11 \cdot 29 \cdot 71 \cdot 911$

(Continued)

Appendix A.2.   (*Continued*)

n	L_n	Prime Factorization of L_n
36	33,385,282	$2 \cdot 7 \cdot 23 \cdot 103681$
37	54,018,521	54018521
38	87,403,803	$3 \cdot 29134601$
39	141,422,324	$2^2 \cdot 79 \cdot 521 \cdot 859$
40	228,826,127	$47 \cdot 1601 \cdot 3041$
41	370,248,451	370248451
42	599,074,578	$2 \cdot 3^2 \cdot 83 \cdot 2811427$
43	969,323,029	$6709 \cdot 144481$
44	1,568,397,607	$7 \cdot 263 \cdot 881 \cdot 967$
45	2,537,720,636	$2^2 \cdot 11 \cdot 19 \cdot 31181 \cdot 541$
46	4,106,118,243	$3 \cdot 4969 \cdot 275449$
47	6,643,838,879	6643838879
48	10,749,957,122	$2 \cdot 769 \cdot 2207 \cdot 3167$
49	17,393,796,001	$29 \cdot 599786069$
50	28,143,753,123	$3 \cdot 41 \cdot 401 \cdot 570601$
51	45,537,549,124	$2^2 \cdot 919 \cdot 3469 \cdot 3571$
52	73,681,302,247	$7 \cdot 103 \cdot 102193207$
53	119,218,851,371	119218851371
54	192,900,153,618	$2 \cdot 3^4 \cdot 107 \cdot 1128427$
55	312,119,004,989	$11^2 \cdot 199 \cdot 331 \cdot 39161$
56	505,019,158,607	$47 \cdot 10745088481$
57	817,138,163,596	$2^2 \cdot 229 \cdot 9349 \cdot 95419$
58	1,322,157,322,203	$3 \cdot 347 \cdot 1270083883$
59	2,139,295,485,799	$709 \cdot 8969 \cdot 336419$
60	3,461,452,808,002	$2 \cdot 7 \cdot 23 \cdot 241 \cdot 2161 \cdot 20641$
61	5,600,748,293,801	5600748293801
62	9,062,201,101,803	$3 \cdot 3020733700601$
63	14,662,949,395,604	$2^2 \cdot 19 \cdot 29 \cdot 211 \cdot 1009 \cdot 31249$
64	23,725,150,497,407	$127 \cdot 186812208641$
65	38,388,099,893,011	$11 \cdot 131 \cdot 521 \cdot 2081 \cdot 24571$
66	62,113,250,390,418	$2 \cdot 3^2 \cdot 43 \cdot 307 \cdot 261399601$
67	100,501,350,283,429	$4021 \cdot 24994118449$
68	162,614,600,673,847	$7 \cdot 23230657239121$
69	263,115,950,957,276	$2^2 \cdot 139 \cdot 461 \cdot 691 \cdot 1485571$
70	425,730,551,631,123	$3 \cdot 41 \cdot 281 \cdot 12317523121$

(*Continued*)

Appendix A.2. (*Continued*)

n	L_n	Prime Factorization of L_n
71	688,846,502,588,399	688846502588399
72	1,114,577,054,219,522	$2 \cdot 47 \cdot 1103 \cdot 10749957121$
73	1,803,423,556,807,921	$151549 \cdot 11899937029$
74	2,918,000,644,027,443	$3 \cdot 11987 \cdot 81143477963$
75	4,721,424,167,835,364	$22 \cdot 11 \cdot 31 \cdot 101 \cdot 151 \cdot 12301 \cdot 18451$
76	7,639,424,778,862,807	$7 \cdot 1091346396980401$
77	12,360,848,946,698,171	$29 \cdot 199 \cdot 229769 \cdot 9321929$
78	20,000,273,725,560,978	$2 \cdot 32 \cdot 90481 \cdot 12280217041$
79	32,361,122,672,259,149	32361122672259149
80	52,361,396,397,820,127	$2207 \cdot 23725145626561$
81	84,722,519,070,079,276	$22 \cdot 19 \cdot 3079 \cdot 5779 \cdot 62650261$
82	137,083,915,467,899,403	$3 \cdot 163 \cdot 800483 \cdot 350207569$
83	221,806,434,537,978,679	$35761381 \cdot 6202401259$
84	358,890,350,005,878,082	$2 \cdot 72 \cdot 23 \cdot 167 \cdot 14503 \cdot 65740583$
85	580,696,784,543,856,761	$11 \cdot 3571 \cdot 1158551 \cdot 12760031$
86	939,587,134,549,734,843	$3 \cdot 313195711516578281$
87	1,520,283,919,093,591,604	$22 \cdot 59 \cdot 349 \cdot 19489 \cdot 947104099$
88	2,459,871,053,643,326,447	$47 \cdot 93058241 \cdot 562418561$
89	3,980,154,972,736,918,051	$179 \cdot 22235502640988369$
90	6,440,026,026,380,244,498	$2 \cdot 33 \cdot 41 \cdot 107 \cdot 2521 \cdot 10783342081$
91	10,420,180,999,117,162,549	$29 \cdot 521 \cdot 689667151970161$
92	16,860,207,025,497,407,047	$7 \cdot 253367 \cdot 9506372193863$
93	27,280,388,024,614,569,596	$22 \cdot 63799 \cdot 3010349 \cdot 35510749$
94	44,140,595,050,111,976,643	$3 \cdot 563 \cdot 5641 \cdot 4632894751907$
95	71,420,983,074,726,546,239	$11 \cdot 191 \cdot 9349 \cdot 41611 \cdot 87382901$
96	115,561,578,124,838,522,882	$2 \cdot 1087 \cdot 4481 \cdot 11862575248703$
97	186,982,561,199,565,069,121	$3299 \cdot 56678557502141579$
98	302,544,139,324,403,592,003	$3 \cdot 281 \cdot 5881 \cdot 61025309469041$
99	489,526,700,523,968,661,124	$22 \cdot 19 \cdot 199 \cdot 991 \cdot 2179 \cdot 9901 \cdot 1513909$
100	792,070,839,848,372,253,127	$7 \cdot 2161 \cdot 9125201 \cdot 5738108801$

Bibliography

[1] P. Odifreddi (translated into English by Arturo Sangalli), *The Mathematical Century: The 30 Greatest Problems of the last 100 Years*, Princeton University Press (2006).

[2] T. Cai, *A Modern Introduction to Classical Number Theory*, World Scientific Press (2021).

[3] L. E. Dickson, *History of the Theory of Numbers, I–III*, Chelsea Publishing Company, New York (1952).

[4] M. Kline, *Mathematical Thought from Ancient to Modern Times* (2 Volumes), Oxford University Press (1990).

[5] G. H. Hardy, *An Introduction to the Theory of Numbers*, Oxford University Press (1972).

[6] F. Cajoria, *A History of Physics*, Maven Books (2019).

[7] T. Koshy, *Fibonacci and Lucas Numbers with Applications*, John Wiley & Sons (2nd Ed., 2017).

[8] P. Ribennoim, *The New Book of Prime Number Records*, Springer-Verlag, New York (3rd Ed., 1996).

[9] K. H. Rosen, *Elementary Number Theory and its Applications*, Addison-Wesley, London (5th Ed., 2004).

Index

www.ingramcontent.com/pod-product-compliance
Lightning Source LLC
Chambersburg PA
CBHW050552190326
41458CB00007B/2009